Miniature Ringbom Engines

James R. Senft

Moriya Press

ISBN 0-9652455-3-5

Printed in the United States of America

Miniature Ringbom Engines

May God grant me to speak as He would wish
and conceive thoughts worthy of the gifts I have received,
since He is both guide to Wisdom and director of sages;
for we are in His hand, yes, ourselves and our sayings,
and all intellectual and all practical knowledge.

Wisdom 7: 15-16

PREFACE

Building miniature engines can be an end in itself. Small scale engines are very often built purely for the satisfaction of it, and not to do anything useful. Some miniature engines are eventually put to "work" driving a model boat or toy merry-go-round or maybe even a fan. But that just adds more enjoyment to the basic delight of seeing the engine run. The first reward of building a miniature engine is in seeing it running.

Among the more entertaining of miniature prime movers are the hot air or Stirling cycle engines. These have been favorites of model makers for well over a century, and have been made in a wide variety of sizes and styles. They are relatively easy to make, safe to run, quiet, and offer endless play for the ingenuity of the mechanically minded.

At the present time, interest in model hot air engines seems to be at an all time high. This is perhaps because of a corresponding interest among engineers and scientists in the full size Stirling engine, which is seen as an environmentally advantageous alternative to the internal combustion engine. The full-size research and development effort has uncovered new and interesting concepts which are ideally suitable for model subjects.

Perhaps the most exciting of these scientific developments centers around a type of Stirling engine known as the Ringbom. The Ringbom engine is a Stirling with a normal crank-coupled power piston but a "free" displacer piston. The displacer is mechanically free of any linkage and is driven back and forth automatically by the changing pressure within the machine. It is an utterly fascinating engine to watch in operation and all the more satisfying to build.

The Ringbom engine offers exceptional scope for miniature engineering projects. Not only can the conventional styles of hot air engines be modeled in Ringbom form, but a whole new world of possible configurations is opened up by our new technical understanding of the Ringbom.

This first book on miniature Ringbom engines is intended as a practical introduction to this new branch of miniature engineering. It is primarily a how-to book, focused on making working engines. Detailed plans and instructions for building three different miniature Ringbom engines are at the core. These range from a tiny flame heated high speed engine to a larger machine which operates for several hours on the thermal energy in a few cups of hot water.

A sound grasp of fundamentals is essential for successful work. The first chapter explains the basic theory of the Ringbom in simple terms, but thorough enough so that the builder will clearly understand the operation of these machines. The construction chapters also contain further details of the principles behind these engines, so that by the time the reader advances to the final chapter, he or she should feel confident about trying out some original ideas. The last chapter presents some interesting Ringbom engine concepts, many of which have not yet been tried by anyone on any scale and which will hopefully spark even more innovations. Sound guidelines for modifying engine designs complete the book and ready the reader for more adventurous projects beyond.

The user of this book is assumed to have a fair working knowledge of a lathe, drill press, and the range of metal working hand tools. But the book caters to readers just emerging from the beginning stages of engineering in miniature by offering detailed discussion of the more challenging operations. And throughout the book, there is plenty of material to meet the needs of those belonging to the "Society of Armchair Engineers" who perhaps are temporarily without the time or facilities to do the work in actuality but enjoy the mental challenge of figuring out clever setups, coming up with alternative methods, and cooking up new engine designs to make someday.

Building miniature engines can also be much more than an end in itself. The book will hopefully be useful to educators and students. The engines described make excellent demonstration devices for the science classroom or laboratory. They would also make new and interesting school shop projects, especially the larger low temperature engine described in Chapter 3. The student who completes one of these engines will have gained an insightful introduction to the physics of heat engines and can use the machine for laboratory experiments and science exhibitions.

The book also will be of interest to the industrial or university researcher. Model engineering does not always follow full size practice. It sometimes shows the way. This is one of the more exciting aspects of this avocation. In the case of Ringbom engines, it is already true that small scale work led the full size. The tiny model engine "Tapper" described in Chapter 2 inspired the development of a rather complete theory of operation for high speed Ringbom engines and led to the construction of several large experimental engines.† For the scientist or engineer, miniature engines are recommended as an excellent way to rapidly gain experience in the field and try out new ideas quickly. In the hands of any good machinist, this introductory book will quickly yield working models for the researcher to ponder for scaling up to practical sizes.

To all the users of this book, the author wishes the very best of the enjoyable calling of miniature Ringbom engines.

James R. Senft
August, 2000

---

† This is covered in detail in the author's technical monograph *Ringbom Stirling Engines*, Oxford University Press, 1993, which is a natural companion volume to this one, especially if interest in developing more of these remarkable machines becomes serious.

Contents

Chapter 1. Ringbom Stirling Engines

Much of the early development of the Stirling hot air engine centered on the problem of finding a good mechanism for driving its piston and displacer. It is an elusive problem. A superior drive system for the piston and displacer must be a judicious compromise or combination of practicality, efficient operation, and motion as close to the thermodynamic ideal as possible. Accordingly, the mechanism problem for Stirling engines has attracted many inventive minds and brought out many ingenious solutions. Among the most clever and elegant is the one discovered by Ossian Ringbom.

The Original Ringbom Engine

In 1907, Ringbom patented a remarkably simple form of the Stirling engine. His invention was for ". . . a hot air engine in which movement of the displacing piston is obtained without the connection of rods or cranks or eccentrics or other mechanical parts . . ."[1] It was a separate cylinder or "gamma" type Stirling engine with no apparent displacer drive mechanism. Aside from the absence of links, levers, or cranks to move the displacer, its most prominent feature was an oversize displacer rod.

In conventional Stirlings, the displacer rod is made as small as possible. It only needs to be large enough to bear the load of driving the displacer back and forth as fast as the mechanism may force it. Anything larger adds unnecessary weight and, more subtly, reduces the pressure-changing effect of the displacer.

In Ringbom's original engine, the displacer rod was large enough in area so that rising internal engine pressure near the end of the compression stroke would lift the displacer . This transferred air from

---

[1] U. S. Patent #856102

the cold to the hot end, as in any Stirling, and the additional pressure rise produced the outward power stroke of the piston. After most of the expansion stroke had taken place, the pressure would become low enough again so that the displacer would drop back down by gravity. At the very end of the outward piston stroke, a small port would be uncovered to replenish any air that was lost through leakage. The cycle could then repeat with flywheel momentum supplying the energy for the compression stroke.

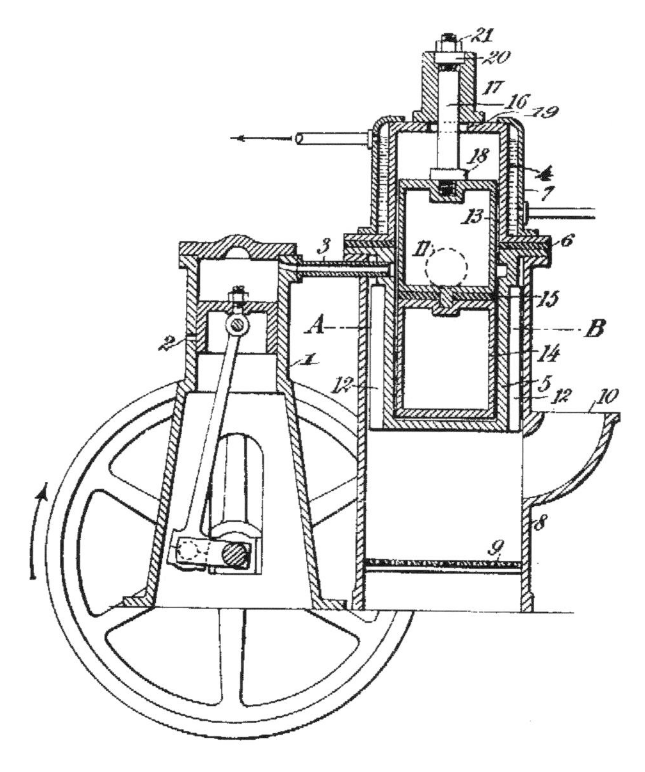

Patent drawing of Ringbom's original engine.

It is not known whether Ringbom actually built and ran his invention. There is no doubt that it would have worked just as he described it in the patent if everything was proportioned just right. It would not have been a fast running engine because it used gravity to return the displacer, but still his engine could have compared quite well with the hot air engines of the time, which all ran at low speeds. And above all the others, it had the advantage of unrivaled mechanical simplicity!

The Modified Ringbom

Relying upon gravity for one stroke of the displacer would limit the top speed of the Ringbom as originally conceived.[2] Only rather recently was it realized that just eliminating Ringbom's "snifter" port at bottom dead center of the piston stroke would remove the gravity speed limitation.[3] Omitting this port forces the mean pressure within the engine to establish itself very nearly equal to the atmospheric pressure outside the engine. This is because of the inevitable leakage past the piston and displacer rod clearances, minimal though it may be. After a brief initial period of cranking and asymmetric running, the average engine pressure will come to equilibrium with external pressure. The engine pressure then will continually fluctuate more or less equally above and below atmospheric pressure. This drives the displacer in both directions and makes higher speeds possible. At high speeds, gravity becomes a negligible influence and the Ringbom engine thus modified can run in any position.

The next figure shows the main features of a typical Ringbom engine without the snifter port. It follows ordinary Stirling engine form, except for the larger displacer rod. The piston and the displacer rod are very close but freely sliding fits. The displacer rod actually functions as a small secondary piston. The two work with each other in a kind of "seesaw" fashion. If the main piston is pushed inward, the displacer rod will move outward. On the other hand, if the main piston is quickly pulled outward, the displacer rod will be pushed inward by

---

[2] The upper limit is inversely proportional to the square root of the stroke.
[3] Cf. *Ringbom Stirling Engines* , Senft, Oxford University Press, 1993.

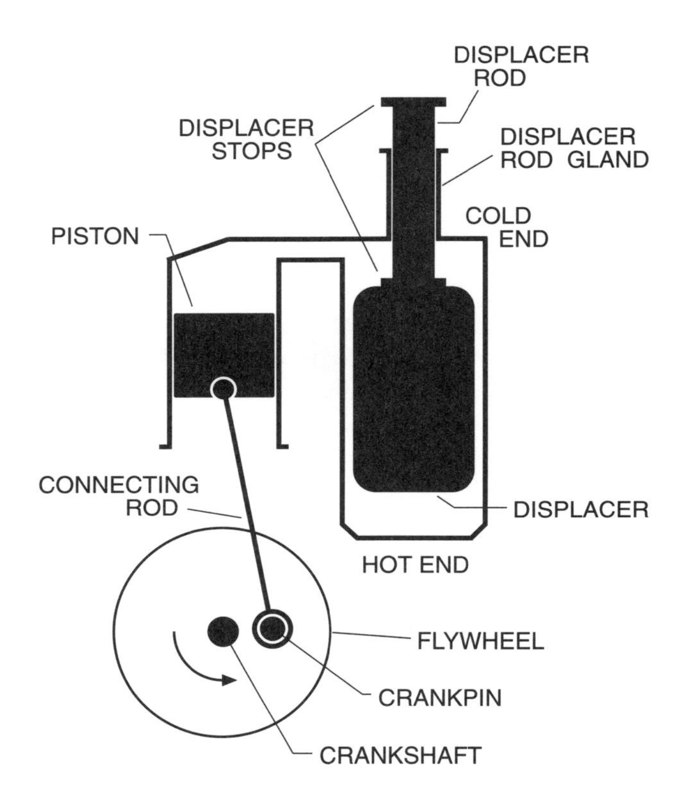

Schematic sectional view of the modified Ringbom.

higher outside pressure. That is how the displacer is made to move. The subtle aspect of the Ringbom is understanding how to get the relation between the piston and displacer motions correct, and how to keep things that way to achieve steady and stable operation.

The Modified Ringbom Cycle

Recall that the ideal Stirling cycle consists of four processes: the compression stroke of the piston, a transfer stroke by the displacer, a piston expansion stroke, and the return transfer stroke of the displacer.[4] These same four processes occur in the cycle of operation of the Ringbom. Part (a) of the next set of figures depicts the compression stroke in progress. The displacer is at rest all the way into the hot end of its chamber. Pressure inside the engine at this point is lower than the external pressure.[5] The piston is being pushed inward, and as it proceeds the engine volume decreases and the pressure inside accordingly increases. The engine is proportioned so that when the piston gets nearer to its inner dead center, the pressure inside the engine equals the outside; the exact point where this occurs is marked by the letter alpha (A) on the flywheel edge in the illustration. The momentum of the flywheel carries the crank past this point, and the further compression takes the pressure inside the engine above the outside pressure. This pushes the displacer outward and initiates the transfer stroke.

In part (b) of the illustration, the piston is moving relatively slowly as it passes through its inner dead center still by flywheel momentum. The displacer is moving outward because of the initial push from the higher pressure inside. As it moves of course, it transfers air from the cold end to the hot. This raises the internal pressure all the more and this effect quickly accelerates the displacer all the way through its stroke. At the end of its stroke, the displacer is rapidly brought to a halt by a resilient cushion or by a dashpot device as described by Ringbom in his patent. The pressure in the engine is now at its peak, just as the piston goes over inner dead center and is ready for its outward expansion stroke.

The pressure inside the engine is above atmospheric during the expansion stroke shown in part (c) until the point marked by omega (Ω) which is exactly half of a revolution from alpha. All during this time the displacer is held stationary against its stop by the elevated pres-

---

[4] Cf. *An Introduction to Stirling Engines*, Senft, Moriya Press, 1993, p. 20ff.

[5] Notice the pressure gauges. At atmospheric pressure, the gauge needle points straight up. It turns to the right for higher pressure and to the left for lower.

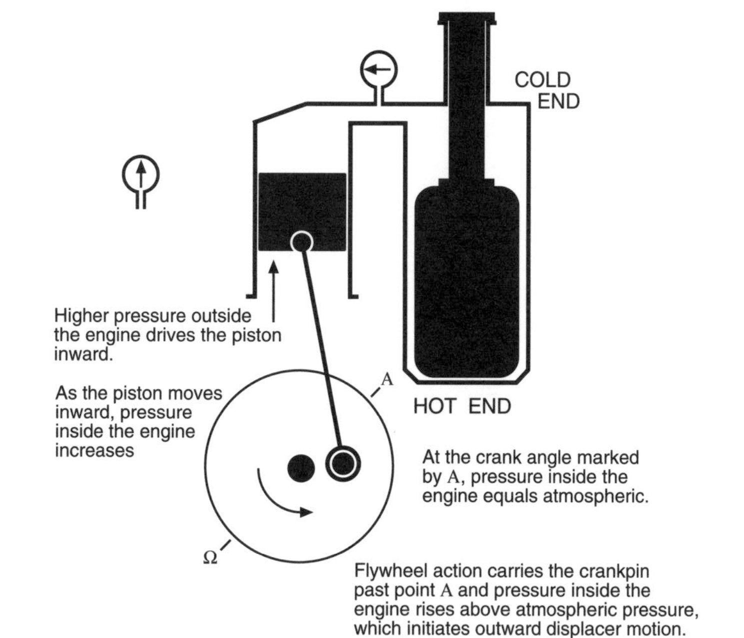

(a) **COMPRESSION STROKE**

(b) **TRANSFER STROKE**

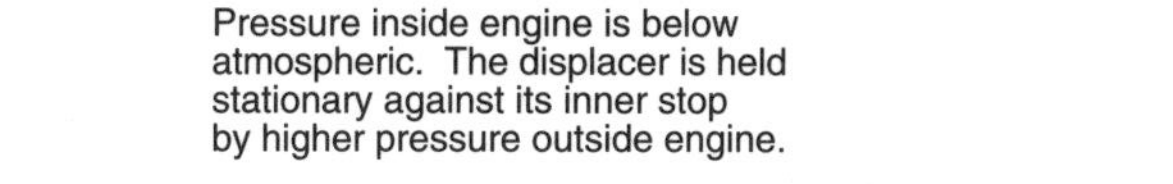

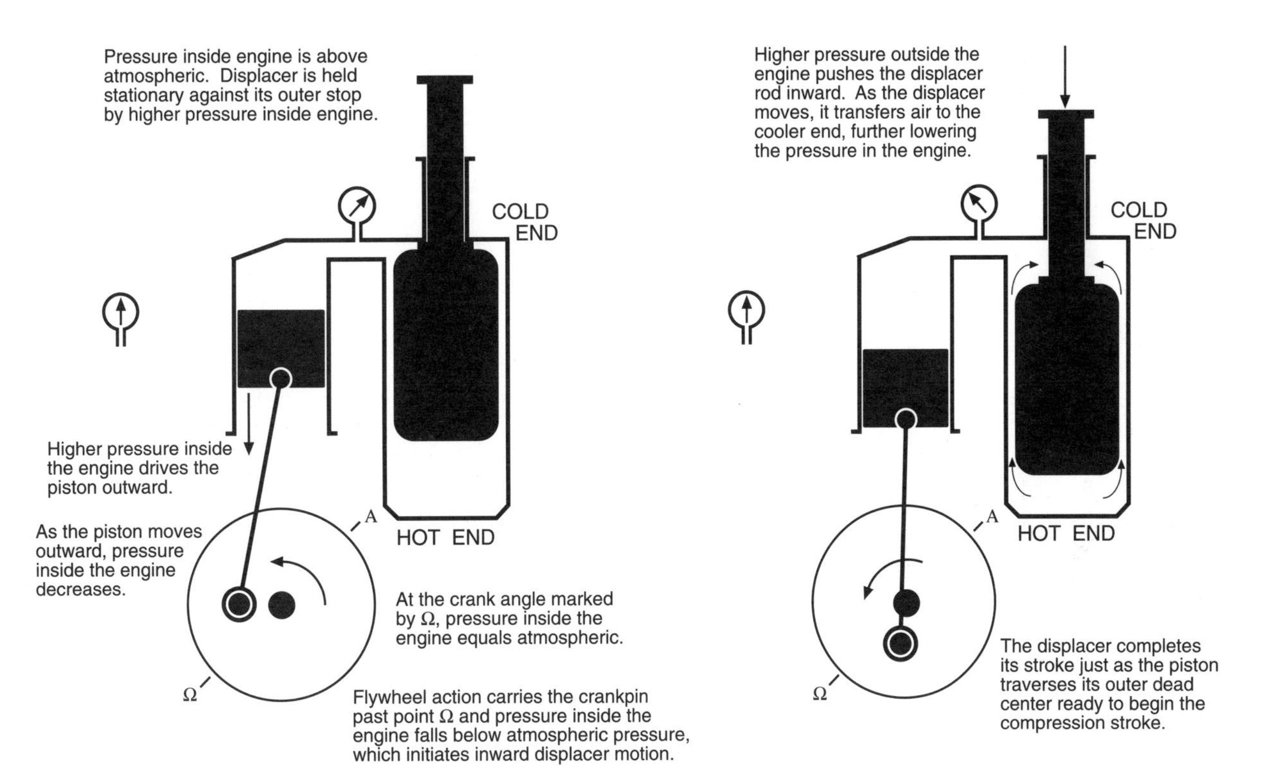

(c) **EXPANSION STROKE**

(d) **TRANSFER STROKE**

sure inside the engine.

The flywheel carries the crank past the omega point, and the additional outward motion of the piston causes the pressure inside to go below atmospheric. This begins the displacer motion inward as shown in (d). This effects the return transfer of displacer chamber air from the hot end to the cold. As in the other transfer stroke, the displacer motion is self-assisting; as the rod moves inward, the displacer body transfers air to the cold side which lowers the engine pressure which drives the displacer rod inward all the more. The engine parts are so proportioned that the displacer completes its stroke and comes to a stop just as the piston passes outer dead center. The compression stroke then begins with the internal pressure at it lowest and the cycle then repeats exactly as before.

Overdriven Mode Operation

At first it might look as though everything would have to be just right to get a Ringbom engine to run. And it might further seem that even if one could be coaxed to run, it would at best be sensitive and temperamental.[6] It turns out however that the modified Ringbom engine as described above is not at all a super critical machine, and will run with as regular a beat as any conventional Stirling engine over a very wide range of speeds. A correctly conceived and constructed Ringbom, if its load is changed, will adjust its cadence in stride and continue to plug away in a stable and steady manner.

This stability is an inherent characteristic of the regime of operation of the modified Ringbom described in the previous section. It has been termed overdriven mode operation to emphasize that the displacer is driven fast enough to complete its stroke before the pressure changes to drive it the other way. When the piston passes the

---

[6] G. Walker related very interesting experience of this in his book *Stirling Cycle Machines*, Clarendon Press, Oxford, 1973, p127-130. This was written long before steady stable overdriven mode Ringbom engine dynamics had been discovered and demonstrated by the engine named Tapper described in the next chapter.

alpha point, the displacer starts to move and completes its stroke before the piston reaches the omega point. Having completed its stroke before omega, the displacer is held stationary by the pressure difference between the inside and outside of the engine. This pressure difference persists, and so the displacer waits, until the piston gets to the omega point. Then the same thing occurs for the return stroke of the displacer. The displacer finishes its stroke before the alpha point and is held in readiness until the alpha point is reached again. This orderly sequence of operation is what we call the overdriven mode.

The dwell periods of the displacer allow the overdriven Ringbom to operate regularly at different speeds. Slowing the engine shaft down by increasing the load will increase the dwell time of the displacer, but the engine still will run in the overdriven mode. Decreasing the load will allow the engine to speed up and this will decrease the displacer dwell time, but as long as there is some dwell time, the engine will continue to run stably in the overdriven mode.

The speed at which the dwell time just becomes zero is the overdriven speed limit. The engine still runs steadily at this speed, but this is the limit. Above this speed, the displacer does not have the time to finish its stroke before the piston has come around and pressure is starting to drive the displacer the other way. The piston gets ahead of the displacer, and the displacer cannot move fast enough to catch up and get back in step. Operation becomes irregular above the overdriven speed limit.

If left to run free, a well-built Ringbom will exceed the overdriven speed limit, but usually not by too much because the operation becomes disordered. Below the overdriven speed limit, operation is rock steady and stable. Up to its overdriven speed limit, a Ringbom behaves much like a conventional Stirling, speeding up when its load is lightened, and slowing down when it is increased. This regular operation persists until the engine is loaded to the point where the flywheel becomes inadequate to keep turning in one direction.

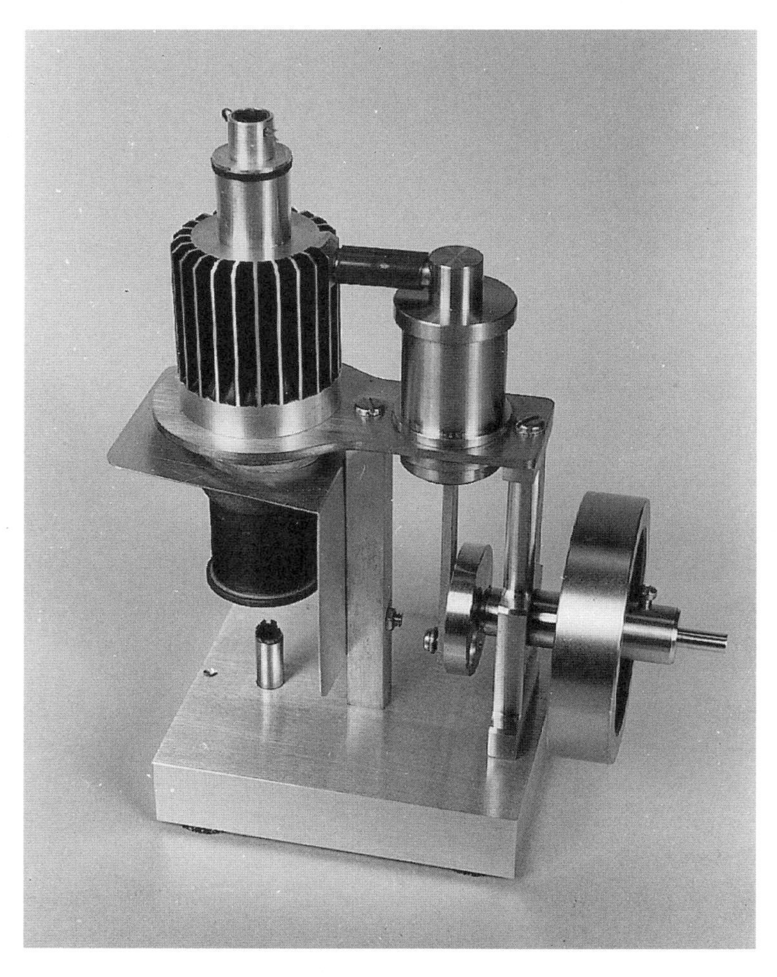

Tapper is a 2cc Ringbom engine that runs fast and steady. This tiny engine was the first to demonstrate that Ringboms are capable of stable high speed operation. As a model, it makes an excellent project and will teach the builder volumes about Ringbom engine dynamics.

Chapter 2. TAPPER

"Tapper" is the most surprising Stirling engine that the author has built. It was begun as most model engine projects are, just for the enjoyment and edification of bringing a new engine to life. Something different however was anticipated from this one because it had a free displacer and there were some new ideas behind its design. I hoped they just might work and give the engine robust high speed operation. And indeed they did work, but I didn't know for sure for quite a while.

The building of Tapper went on for a long time. Construction began in March of 1978. After completing the displacer and the cold and hot ends, the project was interrupted by a long distance move. Then the project languished still boxed up until about a year later. An announced visit by Prof. G. Walker, the reigning expert on Stirling engines, motivated me to unpack the parts and finish the engine before his arrival. The engine described here is the result. If not for the professor's visit, half of the engine might still be in a box somewhere in the shop!

The stable steady behavior of this little engine captivated the distinguished visitor, and inspired two full size engine development projects: a 1 kW Ringbom at Sunpower and a giant 20 kW Ringbom by Walker and Fauvel at the University of Calgary.[1] Now having a much better understanding of Ringbom engines, the little engine is no less, and in some ways even more, fascinating to watch run.

Running Characteristics

Tapper is heated by a tiny alcohol flame, about the size of a birthday candle flame, issuing from a wick tube in the base. About 15 seconds

---

[1] These engines are described in the book *Ringbom Stirling Engines*, Senft, Oxford University Press, 1993.

after lighting up, if the flywheel is nudged into motion, the engine begins to oscillate back and forth. After another 15 seconds, the engine will have warmed up enough go into full stroke operation. After it passes over dead center for the first time, it begins to run with continuous rotary motion in one direction or the other, gradually picking up speed as the hot end continues to increase in temperature.[2] With the small flame usually used, the top speed of the engine is about 900 rpm. With higher flame settings, the engine has reached speeds over 1500 rpm, but this is above the overdriven limit, and operation, though highly entertaining, is erratic.

The engine, like all Ringboms, runs equally well either way! Once started off in one direction, it will tenaciously continue to run in that direction with any load short of stalling it. The inherent stability of its overdriven mode operation can be nicely demonstrated by the "thumb and forefinger brake" applied to the shaft. Under the load of a gentle squeeze, the speed drops but the engine continues to regularly tap along in the same direction. With a harder squeeze, the speed decreases further, but the engine still steadily plugs along. This stability prevails over the entire range of speed down to the stall point, which is about 250 rpm with the small flame. With a larger flywheel, the stall point can be made much lower.

The unidirectional stability with load change that Tapper exhibits is dramatic to see. It is obvious to the most casual observer that the displacer is free of mechanical connection; its upper end can be clearly seen bobbing up and down. And the machine runs in either direction. Yet in spite of this, the engine runs steadily under load. It is perplexing to someone who hasn't read Chapter 1 indeed! Even more startling to the uninitiated is the engine's behavior near the overdriven speed limit. If left to run free of any load, the overdriven limit is exceeded and phasing continually shifts and erratic operation results. The engine runs chaotically, sounding and appearing totally inca-

---

[2] Once in a while, maybe one in a hundred times, the engine may stop right on one of the dead centers, and a second nudge is necessary to start it off and running.

pable of handling the slightest of loads which could be applied. But when a load is applied to the shaft, the speed drops into the over-driven mode range and the engine immediately organizes itself to take to the task with a beat as regular as a conventional Stirling! It is amazing to observe, even if one knows Chapter 1.

A view of Tapper running at high speed. Its heat source is a small alcohol flame which is not quite visible in the photograph.

The engine makes a tapping sound as it runs from which its name is derived. This sound is helpful in observing the stability of the engine's operation, for the human ear is amazingly adept at detecting regular rhythms. It can easily distinguish between regular beats and erratic bouncing. The free action of the displacer can be demonstrated by suddenly holding it down with a finger. The engine will immedi-

TAPPER

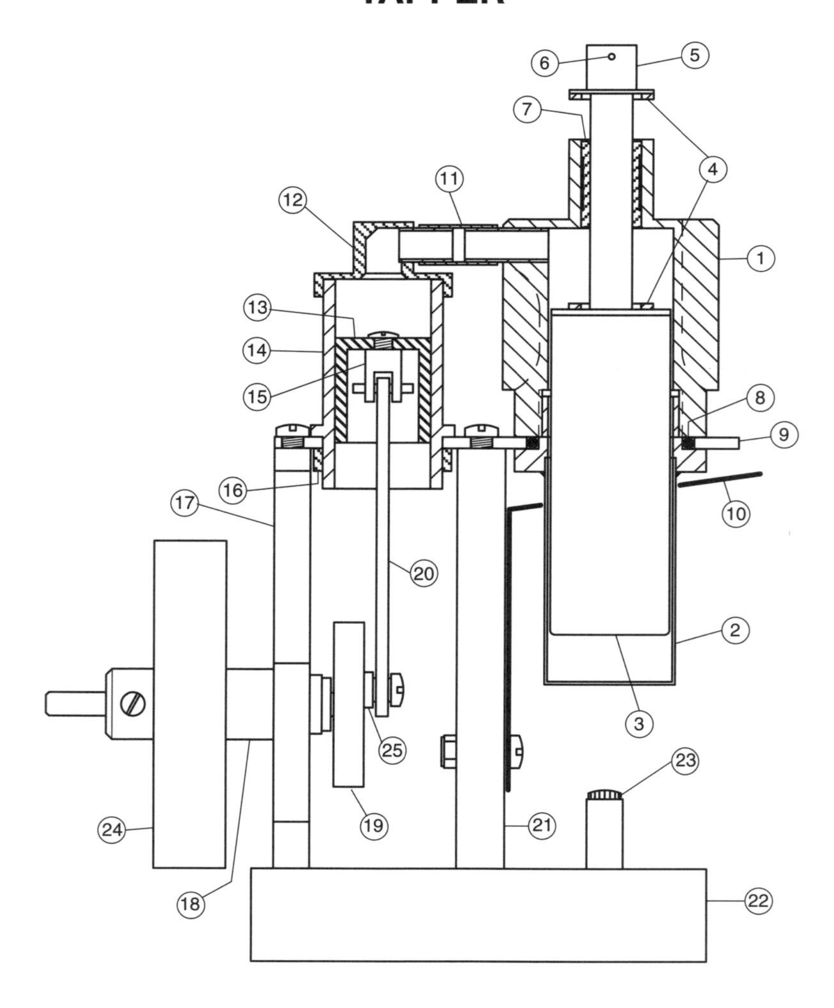

ately begin coasting to a stop. If let go just before completely stopping, the displacer will spring into action again and the engine will steadily climb back up in speed. A finger can also be used to limit the displacer stroke if held rigidly some distance above the cold end. By decreasing the free travel in this way, the speed and power of the engine can be restricted.

Tapper Parts List	
Item No.	Description
①	Cold End
②	Hot End
③	Displacer Assembly
④	Displacer Cushion
⑤	Displacer Stop
⑥	Pin
⑦	Displacer Rod Bushing
⑧	O-Ring
⑨	Cylinder Plate
⑩	Heat Deflector
⑪	Connector Sleeve
⑫	Cylinder Cap
⑬	Piston
⑭	Cylinder
⑮	Piston Yoke
⑯	Cylinder Retaining Ring
⑰	Front Standard
⑱	Bearing Housing
⑲	Crankshaft
⑳	Connecting Rod
㉑	Pillar
㉒	Base
㉓	Wick
㉔	Flywheel
㉕	Crankpin

Building Tapper

The delight of building another Tapper and seeing it spring to life, watching its antics, and learning its personality first hand await anyone who can handle a lathe and drill press. For the most part, standard materials and methods are used with which the reader is assumed familiar. If you have successfully built a conventional hot air engine or two, then you are especially well prepared for making a Ringbom. If not, then the best preparation is to build a conventional kinematic Stirling. Many hot air engine construction articles have been published in model engineering and other magazines over the years.

If you have not yet put together a conventional Stirling and are anxious to have a go at Tapper, you must keep in mind that generally speaking, miniature hot air engines are rather delicate machines.[3] Their power-to-size ratio is minuscule compared to steam or internal combustion engines, and therefore they must be very well yet freely fitted in order to function at all. Any binding in the mechanism or air leaks seriously affect performance and easily can prevent an engine from running altogether. The piston fit is especially important. The piston must travel in the cylinder without any stiffness or drag, and yet be close fitting to give virtually no leakage. The drag of piston rings is too much for tiny model hot air engines. An accurate close mating fit is critical for good operation.

There are two close free fits necessary in Ringbom models. In addition to the piston, the displacer rod must be an equally good fit in its bushing. In a conventional Stirling where the displacer is driven through linkage from the crankshaft, a little extra stiffness in its movement is of course a power loss, but only a direct one. Power to move the displacer equals power taken from the engine. The engine may consume a little more engine power moving a stiffer than normal dis-

---

[3] "Hot Air Engine Wrinkles" Prof. Dennis Chaddock, *Model Engineer*, 1980, Vol. 146, No. 3644, p.1333-4.

placer, but may still run decently. But in a Ringbom, any extra friction on the displacer rod slows down its motion relative to the piston. That retards the phasing, and that further reduces power output and causes erratic operation to occur at lower speeds. In general, any defect in a Ringbom engine which affects displacer motion results in an amplified engine power loss. All these considerations become more critical as size decreases.

Power Cylinder and Piston

Tapper is a small engine. As can be seen in the drawings, the power cylinder bore is 1/2" and the stroke is 5/8". This gives Tapper a swept volume of 2 cc. The cylinder on the original Tapper was turned from brass bar, and the interior bored and lapped to close limits of geometric truth and a fine surface finish.[4]

The piston in Tapper is graphite, carefully turned from solid rod to a close fit in the cylinder. The self-lubricating quality of the graphite piston permits it to run dry in the cylinder, without the need for any oil. This avoids the viscous drag losses associated with liquid film lubrication. In tiny flea-power engines, viscous oil drag can be a major loss item. Despite the softness of graphite rod, it gives good life in these lightly loaded engines provided the cylinder surface is smooth and the connecting rod is relatively long. Graphite pistons have been used on a number of model engines up to a 2" bore Ericsson with every satisfaction. Many other piston and cylinder materials have been used on miniature Stirling engines with good degrees of success, but this author has used graphite exclusively on all miniature Ringbom engines.

A generous wall thickness is recommended for graphite pistons because of the relative fragility of the material. A mirror finish in the

---

[4] For more information on cylinders and pistons for Stirling engines, cf. the following articles: "Fitting Cylinders and Pistons", Senft, *Live Steam Magazine*, Sept. 1979, p.14-16 and "A Short Course in Lapping" *Ibid.*, Oct. 1979, p.26-29.

cylinder is not essential, but the smoother the better. Initially, there will be some rapid abrading of the graphite piston by the roughness of the cylinder wall. No matter how smooth the cylinder is, some of this will occur. For this reason it is a good idea to turn the piston to a slightly stiff fit into the cylinder and work it back and forth in the cylinder until the fit becomes free. It should only take a dozen strokes or so; any more than this means the piston is just too big. This process packs the low areas of the cylinder surface with graphite particles. The cylinder in effect becomes smoother and wear slows down to an acceptable rate. If the cylinder is made too rough, the larger pockets do not pack up or hold the packing particles as well. Abraded graphite leaves the pockets and is pushed along to the ends of the cylinder as the piston reciprocates, and fresh abrasion takes place. The result is that wear is more rapid, so aim for as smooth a bore as you can obtain.

Graphite rod is used by welders to temporarily plug up holes near where welding is being done to keep them clear. I have used this rod on several engines with good results. However the graphite rods that are sold in welding supply shops are not always of a good grade and some contain clay or other material which causes excessive cylinder wear. Highly pure grades of graphite are used for electrodes in electro-discharge machining (EDM) and are available from industrial suppliers. EDM grade graphite is excellent piston material and easy to machine.

Chucked in the lathe, the piston stock for Tapper was drilled and counterbored to leave a 1/16" wall when finished. Then the outside was turned to fit the finished cylinder. For an extra fine infeed for turning to final size, the lathe topslide can be set to just under six degrees to the longitudinal axis. You can check on your calculator that $\sin(5.74°) = 0.1$. If you remember your trigonometry this means that if the topslide is set over to this angle, every thou fed through the topslide feed screw will move the tool in toward the lathe axis by a tenth. This gives very fine control of final cuts. As mentioned above, it is a good idea to turn a graphite piston to a slightly snug initial fit, but not tight. The finished piston was parted off the parent stock with a sharp tool to leave a good flat surface on the piston top.

Cylinder Assembly

The cylinder cap was turned from brass. It was fastened to the cylinder with retaining compound, but epoxy or soft solder could also be used. A short length of 3/16" thin wall brass tubing was likewise fastened into the port drilled in the top projection of the cap to make the connection to the cold end of the displacer cylinder section. Upon final assembly, a flexible tube makes the connection between the two cylinders. The cylinder was fastened to the cylinder plate with retaining compound. To increase the bond area, a thin ring was be made to slip over the cylinder end from below as shown on the assembly drawing, and the whole assembly secured with retaining compound. If solder or epoxy is used instead, the ring is not necessary.

Crankshaft and Crankpin

The crankdisc was turned from brass bar stock, but steel is a good choice also. It was pressed onto a length of 1/8" diameter stainless steel rod. Drill rod is also good choice because of its straightness. The disc was tapped to take a small screw at the crankthrow radius of 5/16". The crankpin is a shouldered bushing turned from mild steel. The big end seats against the crankdisc keeping it square to the disc and spacing the connecting rod to prevent it from rubbing the crankdisc. The head of the screw which holds the crankpin to the disc retains the connecting rod as well. The crankdisc was drilled on either side of the crankpin to produce a counterbalancing effect.

Main Bearings and Front Standard

The crankshaft runs in ball bearings. The engine would run with plain bearings, but definitely not as well. If you want to get the full antics of which the engine is capable, it is important to use low friction bearings here. The ball bearings used on Tapper are of the flanged type and simply slip into the ends of the bearing tube which has a 1/4" diameter hole reamed through. The outside of the bearing tube has a reduced diameter boss at the crankdisc end which is press-fitted into the bearing standard. If your bearings are larger on the O.D., the bearing tube diameter will have to be increased. The front standard was turned to profile from a strip of 3/16" thick brass held in the 4-jaw chuck and supported by the tailstock center.

Connecting Rod and Piston Yoke

The connecting rod is cut from 1/16" thick brass sheet. It would really be a good idea to fit a ball bearing on the big end of the connecting rod, but the original Tapper just had just a plain reamed hole riding on the steel crankpin. The little end is pivoted to the yoke with a small steel pin. The piston yoke is aluminum for light weight, but could also be made in brass or steel if balance is not a major concern. Soft washers on either side of the piston head seal the center hole and prevent fracturing the graphite when tightening the screw.

Base and Pillar

The engine base was made from 1/2" thick aluminum. Two holes are drilled and countersunk through to take the screws going up into the bearing standard and the pillar which support the cylinder plate. The base is bored from below to form a cavity for the alcohol fuel. A thin disc of aluminum sheet is epoxied in place to close up the reservoir. The wick tube is 3/16" O.D. thin wall stainless steel tubing for minimal heat conduction to the base. The wick is soft cotton string. For filling, a hole 1/16" in diameter was drilled into the reservoir at an angle from the top rear of the base. The tank is filled from a syringe with a blunt needle inserted into the hole. Filling this way is very neat. The filler hole is visible in the photographs. As when using any liquid fuel, be sure to always dry off any spillage anywhere and cap up and put away the main container before lighting the burner. Also take care not to tip over the engine when in operation.

Displacer Assembly

The displacer for any Ringbom engine must be made with care. It must be made as light as possible to obtain high speed operation. It must be true enough not to rub in its cylinder. Its rod must be geometrically accurate for a superb airtight fit in its bushing and it must be smooth for low friction free movement. Naturally one aims for these conditions in any Stirling engine. But in a kinematic Stirling, one does not have to achieve all of these conditions perfectly to end up with a decently running machine. For example, a little extra displacer mass in a conventional Stirling will not noticeably affect performance, but in a Ringbom it directly sets back displacer phasing, and this in turn limits speed and power.[5] Again, in a Ringbom, any defect in its displacer will have amplified effects on engine performance. The effect of a rod that is a little too loose in its bushing is the same. So is friction, as was mentioned earlier. A Ringbom engine requires all of these conditions to be met to as high a degree of perfection as possible.

It seems that all true miniature Stirling engine enthusiasts save every deep drawn metal can that crosses their path: cigar tubes, condenser cans, medicine containers, lipstick tubes, etc.. In the case of Tapper, the aluminum body of a felt tip marker was just right. After cutting to length just under 1-3/4" and removing the insides, it weighed a mere 2.7 grams which felt impressively light. Looking back, the displacer could have been made a full gram lighter by boring out to reduce the wall to a thickness of .015" . This thickness would still be plenty strong enough. A wooden block bored out to take

---

[5] The overdriven speed limit of a Ringbom is inversely proportional to the square root of its displacer mass, all else being the same.

the cap as a push fit would be necessary to hold and support it for boring, but the extra work would have been well worthwhile. If a suitable drawn cap is not available, then the displacer end can be turned out of solid 5/8" diameter aluminum alloy bar stock - an advantage in making a small engine. The best material for displacers is stainless steel because it resists corrosion and is a relatively poor conductor of heat. But in a model where efficiency is not a primary concern and light weight is, aluminum works pretty well.

As far as strength is concerned, the displacer endpiece can likewise have a thinner wall than the drawing specifies - say .020" overall. But here the weight savings would not justify much extra work since the end as made in the original is only 0.8 gram. The end should actually be made after the rod is finished so that the bore can be made to a push fit on the rod to ensure concentricity.

The rod on Tapper was made of brass because thin wall brass tubing of 7/32" outside diameter was available. The outside was lapped to a good finish with a simple split lap and non-embedding compound.[6] If good straight round tubing is available, then polishing with #400 silicon carbide paper might be sufficient.

The coefficient of thermal expansion of brass is relatively large, and if Tapper is allowed to get too hot the rod grows a little too tight in its graphite bushing. Therefore the drawing calls for a steel rod which should eliminate the problem. But since the rod, like the piston, runs absolutely dry, rusting may be a problem. Stainless steel is the preferred material because of this, but is it more difficult to lap because it has a tendency to gall. However, with plenty of lubrication and light pressure, satisfactory results can be obtained. The displacer rod on the original Tapper weighed in at 2.2 grams.

The three pieces were assembled with retaining compound which has held up well so far. But this little machine does get hot and high temperatures shorten the life of these adhesives. Several 1/32" diameter pins through the parts would be a good idea for added security.

---

[6] Op. cit., *Live Steam Magazine.*

I have done this on larger engines both of the Ringbom and the conventional type. In whatever way the displacer is held together mechanically, it is important to use some sort of sealant in the joints to make sure the displacer is airtight. Epoxy is also used by many constructors to assemble and seal displacers.

The displacer stop was bored to a light push fit on the displacer rod and is held in place by a 1/32" brass cotter pin. A small taper pin or very fine threads would also be good choices for securing these parts. Soft rubber washers 1/32" thick were cemented to the stop and the displacer end as shown in the engine assembly drawing. These are absolutely essential for cushioning the displacer contacts which are quite forceful at high engine speeds. Use the softest rubber available and keep the section small. Soft o-rings glued in place might also work well.

Cold End

The cold end is machined from aluminum rod. The boring of the main cavity, the displacer gland, and the threading should be carried out in the same chucking so that these features are concentric. Concentricity is important to make sure that the displacer will not rub anywhere. In addition, a high level of concentricity makes the gap around the displacer uniform which evenly distributes the flow around the displacer and makes best use of all the area available for heat transfer. The radial gap between the displacer and its cylinder specified for this engine is .015". The gap is not critical unless one is interested in getting the utmost power. The engine should run nicely with a larger gap. Perhaps with a little less power, but not noticeably so when just running the engine free for demonstrations. A smaller engine I built has a radial gap of .018" and a larger one a gap of .030" and both run fast and free. Generally speaking, a smaller gap has higher heat transfer rates but higher fluid flow friction also. Enlarging the gap reduces fluid friction loss but also decreases heat transfer. The optimum is a matter for experimentation, but judging from ordinary running experience, a gap between .015" and .025" for Tapper will give all around satisfactory performance.

The cooling fins on the outside of the cold end were formed through the use of a 1/16" wide Woodruff cutter. The turned, bored, and threaded end was mounted on a mandrel held in a dividing head. Two cuts are necessary to fully form each space, one cut for each side of every fin. Three of the fins are short and terminate in a solid section at the top of the cold end. The drawing shows this section as drilled through 3/16" diameter to take a short length of thin wall tubing to connect up to the power cylinder. As can be seen, there is not much material to spare here, so it is recommended that the end be drilled through only 5/32" diameter and then counterdrilled 3/16" diameter to a depth of 1/8" or 5/32" to accept the connecting tube. This is secured in place with epoxy or retaining compound. The fins shown on the drawing have been increased in width by 50% over the original Tapper for cooler running. Calculations indicate that the temperature should be on the order of 40°F lower with the wider fins.

Displacer Rod Bushing

The displacer rod bushing is made from graphite rod. The hole is bored to a close running fit on the displacer rod. The same techniques as described above for turning the piston apply here. The outside surface of the bushing is turned with a shallow recess. Upon assembly, epoxy is applied to fill the recess as the bushing is slipped into the cold end. The idea here is that the epoxy, although unable to adhere to the graphite, will adhere to the aluminum surface, and will lock the recessed section in place.

Hot End

The simplest way of making the hot end is in two pieces. The stainless steel hot cap can be turned from solid and silver brazed to the turned and threaded ring as specified in the drawing. On the prototype, the threaded ring is brass. The cap is ordinary mild steel, which works well enough, but will rust in time. Tubing was used for this part with a bronze cap turned up and brazed on the end, but turning the whole cap from solid is probably just as easy in the end. In any event, stainless steel would be the best for this. The threaded ring should have a recess accurately counterbored in the end to take the cap. This will ensure good alignment of the joined parts. The counterbore can be made after threading, boring through, and parting off with the part wrung onto a stub mandrel.

Cylinder Plate

The cylinder plate is bored to accept the power cylinder and the displacer cylinder as shown in the drawings. The larger bore must just accept the outside of an o-ring fitting over the hot end spigot. The o-ring should be slightly proud of the surface of the plate so that it will be squeezed down a bit on assembly. This makes the gas-tight seal between the hot and cold ends which clamp tight against either side of the plate.

Heat Deflector

A heat deflector bent up from thin aluminum sheet is fitted to Tapper as shown in the assembly drawing. This keeps the rising combustion gases from impinging against the cylinder plate and unnecessarily heating up the rest of the engine. Even in the simple form used on the prototype it is very effective, but could be improved by adding side sheets and ceramic insulation inside. There should be ample clearance between the top of the deflector and the bottom of the engine plate for ambient air circulation to help cool the upper part.

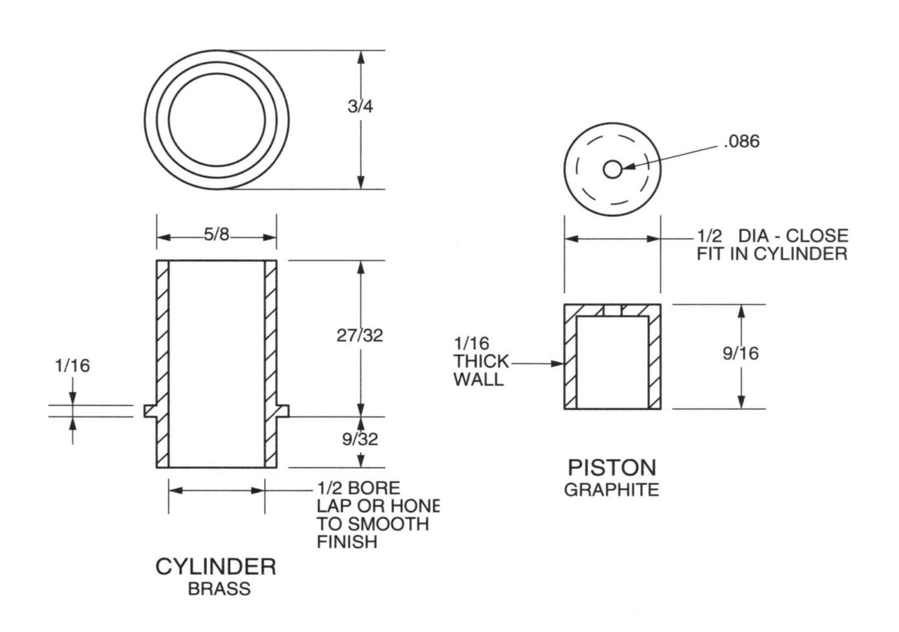
3/4
.086
5/8
1/2 DIA - CLOSE
FIT IN CYLINDER
27/32
1/16
1/16
THICK
WALL
9/16
9/32
1/2 BORE
LAP OR HONE
TO SMOOTH
FINISH
PISTON
GRAPHITE
CYLINDER
BRASS

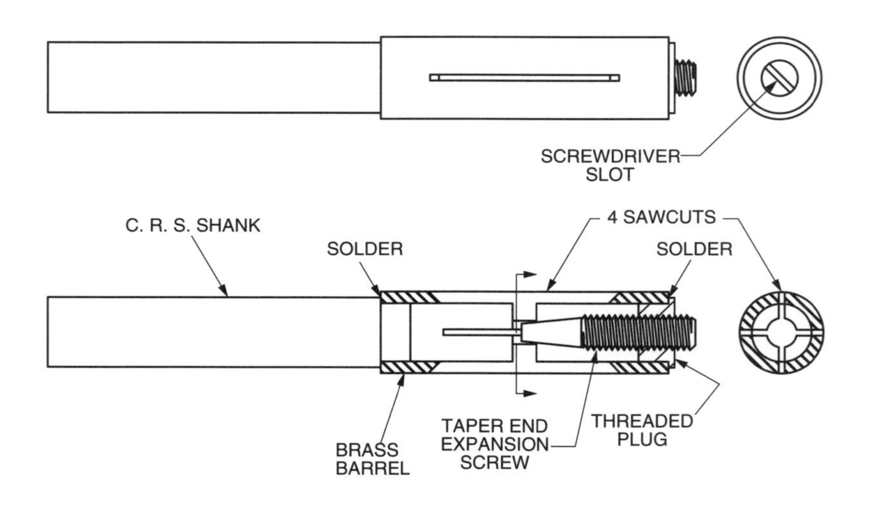
SCREWDRIVER
SLOT
C. R. S. SHANK
4 SAWCUTS
SOLDER
SOLDER
TAPER END
EXPANSION
SCREW
THREADED
PLUG
BRASS
BARREL

BARREL LAP

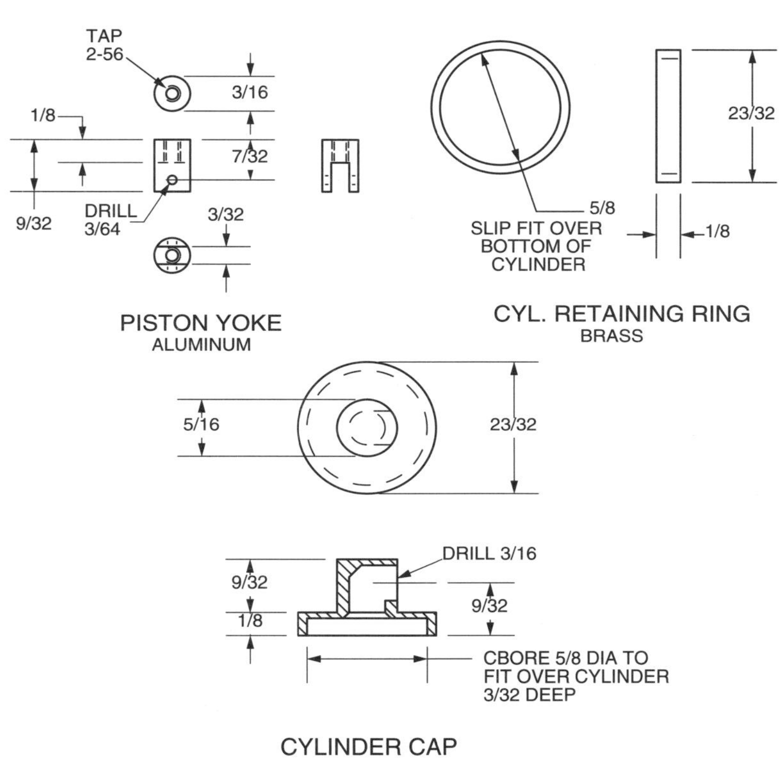

PISTON YOKE
ALUMINUM

CYL. RETAINING RING
BRASS

CYLINDER CAP
BRASS

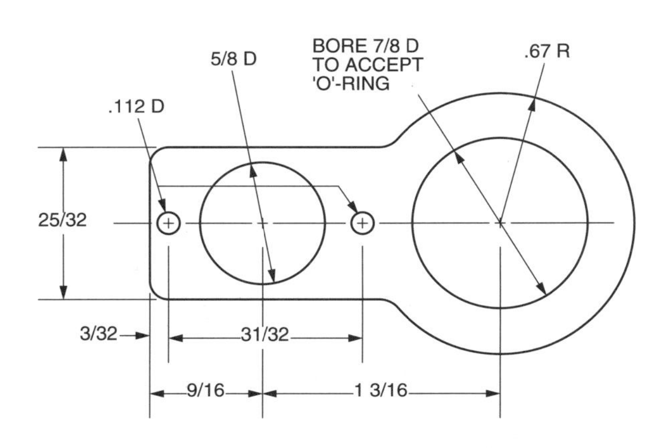

CYLINDER PLATE
1/16 THICK BRASS OR ALUMINUM

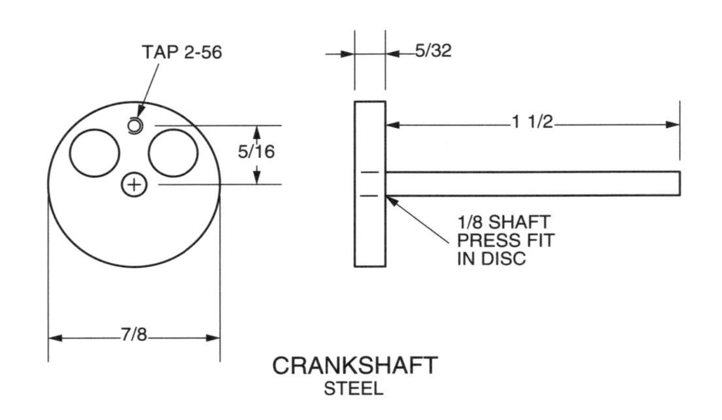
TAP 2-56
5/32
1 1/2
5/16
1/8 SHAFT
PRESS FIT
IN DISC
7/8
CRANKSHAFT
STEEL

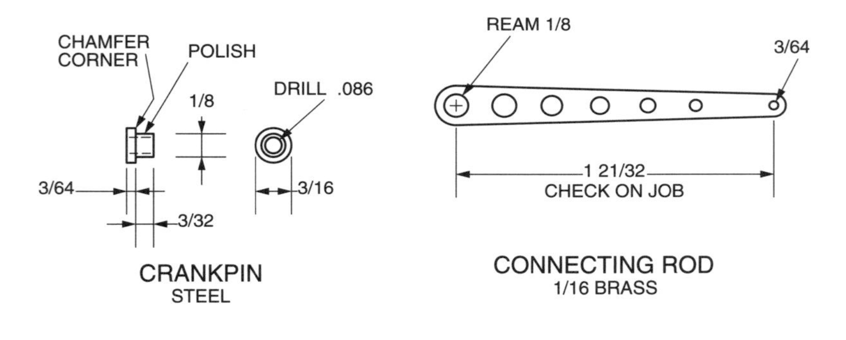
CHAMFER
CORNER
POLISH
1/8
DRILL .086
3/64
3/16
3/32
CRANKPIN
STEEL
REAM 1/8
3/64
1 21/32
CHECK ON JOB
CONNECTING ROD
1/16 BRASS

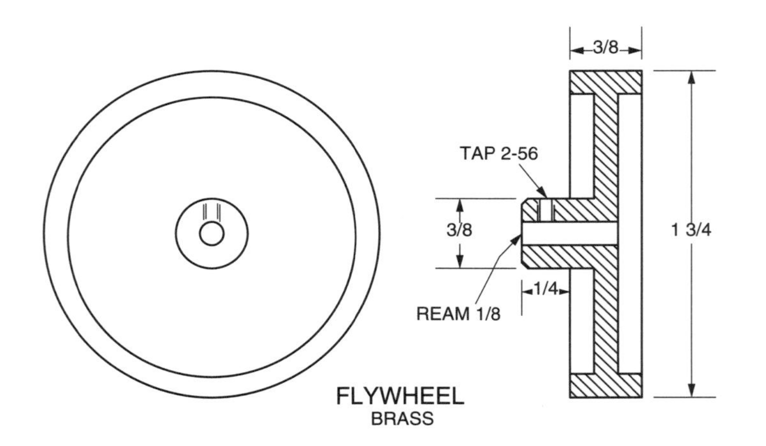
3/8
TAP 2-56
3/8
1 3/4
1/4
REAM 1/8
FLYWHEEL
BRASS

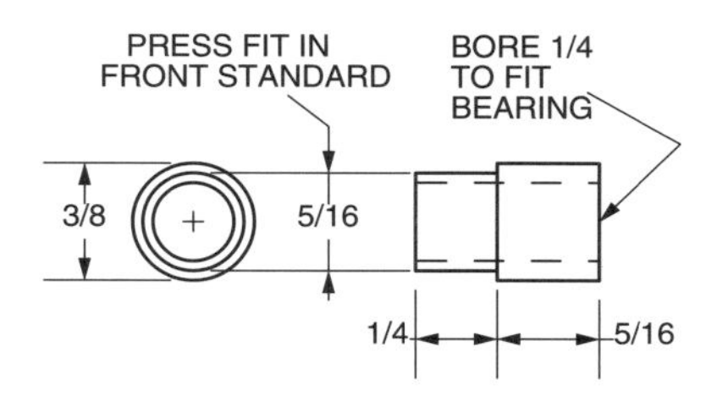

BEARING HOUSING
BRASS

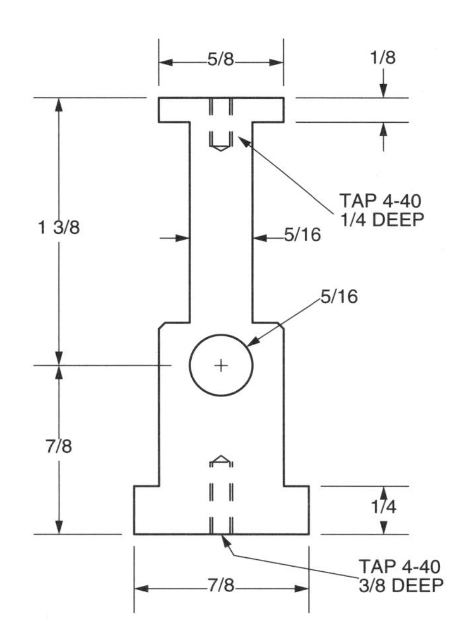

FRONT STANDARD
3/16 THICK BRASS OR ALUMINUM

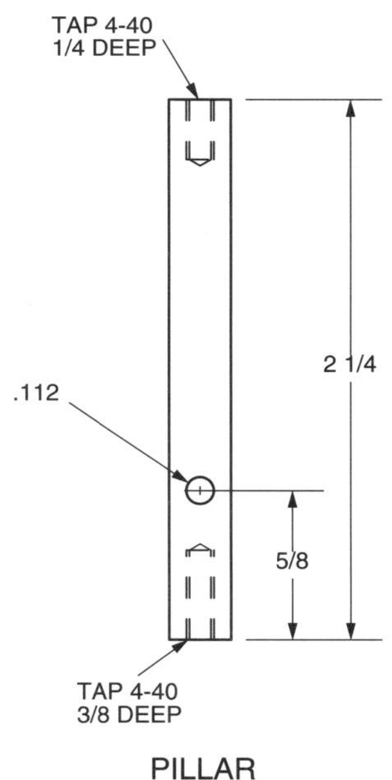

PILLAR
1/4 SQ. BRASS

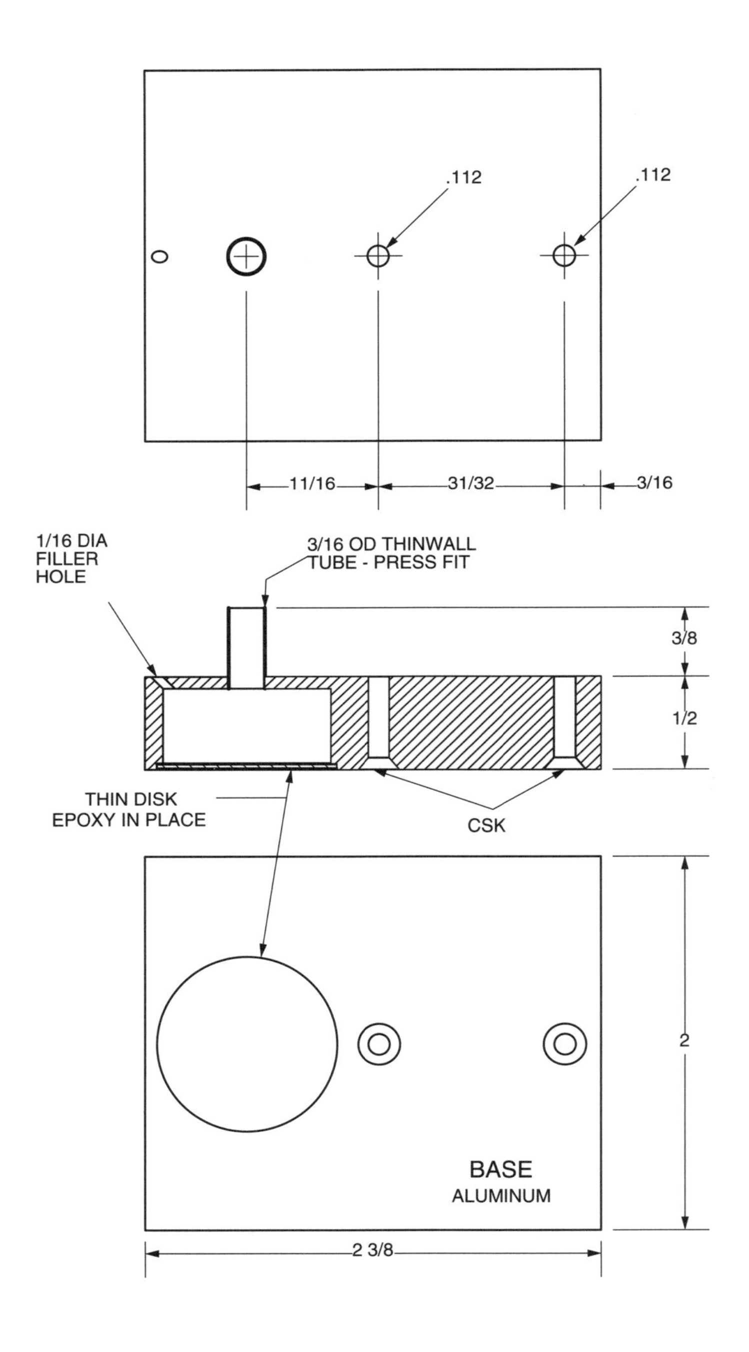
.112
.112
11/16
31/32
3/16
1/16 DIA
FILLER
HOLE
3/16 OD THINWALL
TUBE - PRESS FIT
3/8
1/2
THIN DISK
EPOXY IN PLACE
CSK
2
BASE
ALUMINUM
2 3/8

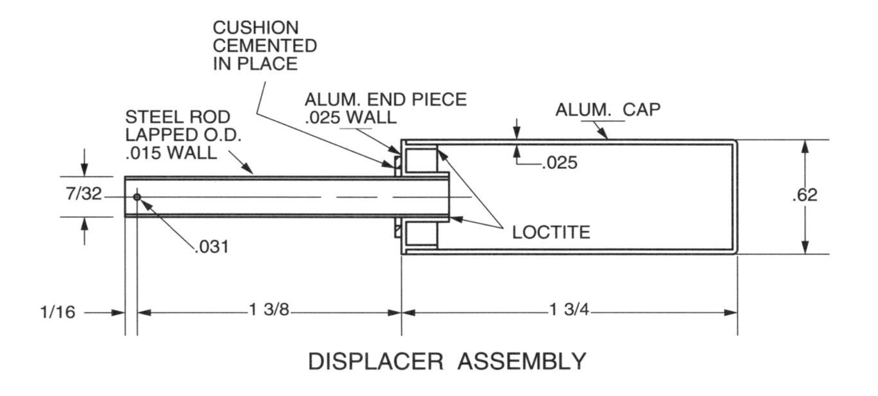

DISPLACER ASSEMBLY

7/16
1/32
5/16

DISPLACER CUSHION
RUBBER - 2 REQ.

CUSHION
CEMENTED
IN PLACE
.020 WALL
BORE TO
FIT ROD
7/16
.031
.20

DISPLACER STOP
ALUMINUM

1.0
THREAD
40 TPI
3/4
9/32
3/64
1/8
SILVER
BRAZE
.65
1 1/4
ST.
STEEL
CAP
.015
.68

HOT END

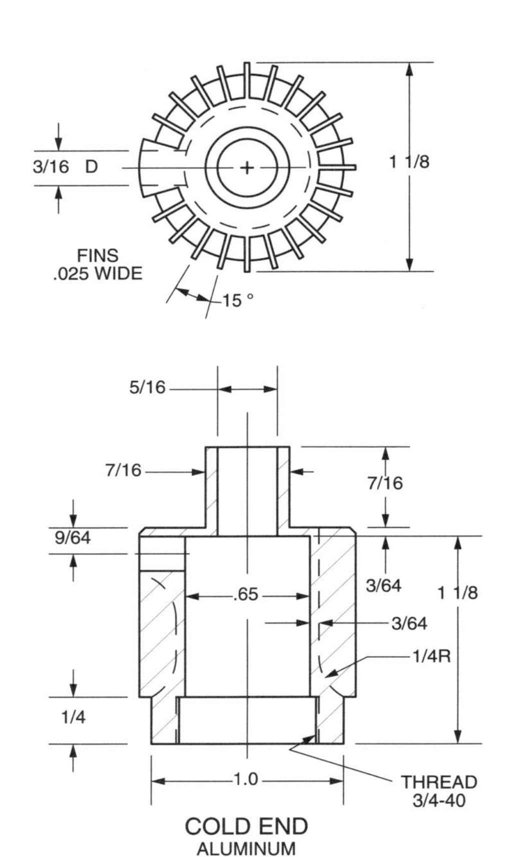

COLD END
ALUMINUM

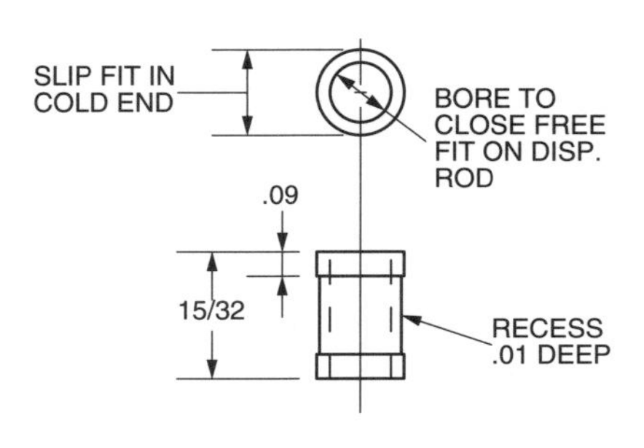

DISP. ROD BUSHING
GRAPHITE

The holes in the Cylinder Plate were machined with a boring head on the milling machine. The part was fastened to a block of wood by four screws and washers around its periphery. This operation could also be carried out in the lathe with the wood bolted to a faceplate or held in a 4-jaw chuck.

The profile of the Front Standard was machined on the lathe.

Final Assembly

The engine must be virtually airtight yet free. A quick leak anywhere or tight spot in the mechanism will doom the engine. The displacer must promptly respond to piston movements. If the crankshaft is rotated a quarter turn on the compression stroke, the displacer should quickly pop up to the top of its stroke. If the displacer action is not lively, performance will suffer. Piston compression must be 'bouncy' when the displacer is held down and the flywheel is flipped over. Some slow leakage past the piston and displacer rod is acceptable and normal, but it must be very slow. In Tapper now, after a lot of running time, if the crank is rapidly turned from bottom to top dead center and held there, it takes about 6 seconds to lose pressure and see the displacer start to fall back down.

Variations

Of course it is not essential to adhere to every detail shown in the drawings and described above. Constructors will no doubt want to change certain features to suit the materials and tools on hand as well as individual tastes and preferences. Engineering in any scale and for whatever purpose is a highly creative endeavor and individuals show their own technical personality in their work. Some stress aesthetics while others put emphasis on performance. Tapper has a lot of scope for all engineering tastes. There is much room in particular for experimentation to improve operation. However, to get the engine running initially, it is a good idea to adhere closely to the parameters that produce the operation of the original Tapper. These are very few in number and are given in the accompanying table. Any engine with these specifications is dynamically similar to Tapper and so will run the same way, given that the fits are good and the friction is low. Of course, a lighter displacer will give even better performance.

Basic Specifications of Tapper
Power piston swept volume: 0.123 in^3 (2cc)
Displacer diameter : 5/8"
Displacer stroke : 5/8"
Displacer rod diameter : 7/32"
Displacer annular gap : 0.015"
Displacer mass : 6 grams

A few years after Tapper was completed, the author sent preliminary versions of the drawings to Dr. Rod Fauvel at the University of Calgary who then produced a very interesting version. Instead of cooling fins, a water jacket surrounds the cold end. This is of the evaporative type with an open top as used on old stationary gasoline engines. Evaporative cooling is based on the fact of nature that water open to the atmosphere cannot be heated above 212°F (100°C). At this temperature (with slight variations for differing elevations) water boils away carrying a lot of thermal energy off with it. The more heat that is applied, the faster the rate of boiling, but the temperature is fixed. Generally speaking, evaporative cooling is very effective because the heat transfer between metal and boiling water is very high.

Another variation in Fauvel's engine are the conical ends on the displacer. This shape gives greater strength and fatigue resistance. The angle is 118° chosen to match the drill used to hollow out the inside. Fauvel reports that this is a good angle for improving the stiffness of the ends without excessively lengthening the parts as would a spherical shape for example. Porting of the power cylinder is also good with this angle and so it was used on the power piston as well.

The sectional drawing shows the form of displacer stops used by Fauvel. The cold end was drilled in several places to accept short lengths of rubber o-ring cord. These are trimmed to stand proud of the surfaces and so contact and cushion the displacer at the ends of its stroke. These displacer cushions are very effective. Fauvel subjected his little engine to extensive testing and used slow motion photography to view and confirm its overdriven mode operation.[7]

A water-cooled version of Tapper by Rod Fauvel.

---

[7] Op. cit., *Ringbom Stirling Engines*, Chapter 7.

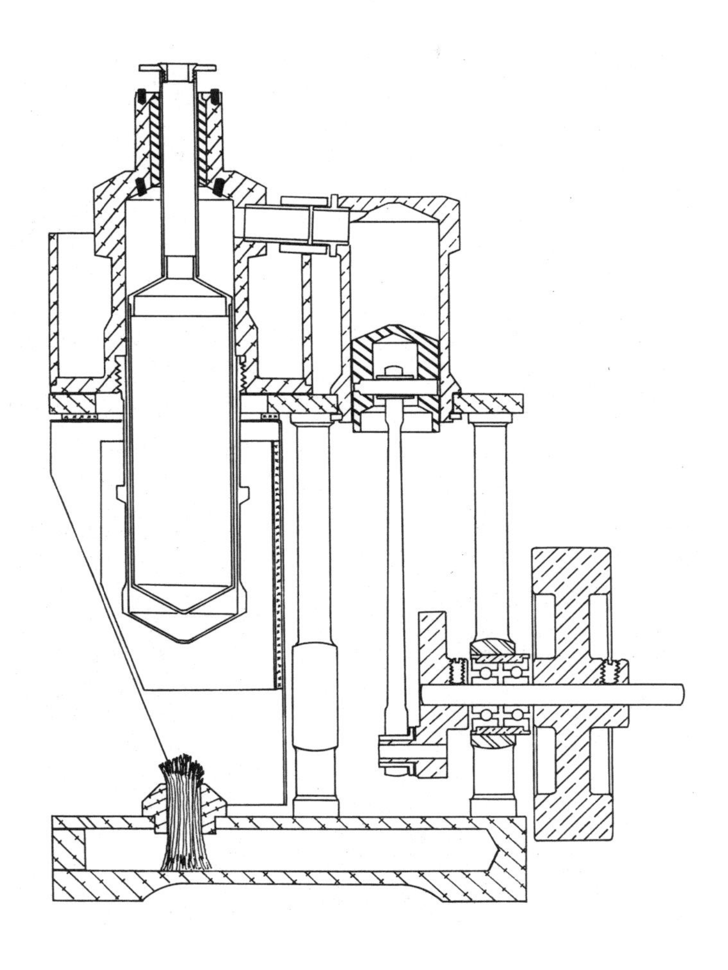

A cross-sectional view of Prof. Fauvel's version of Tapper.

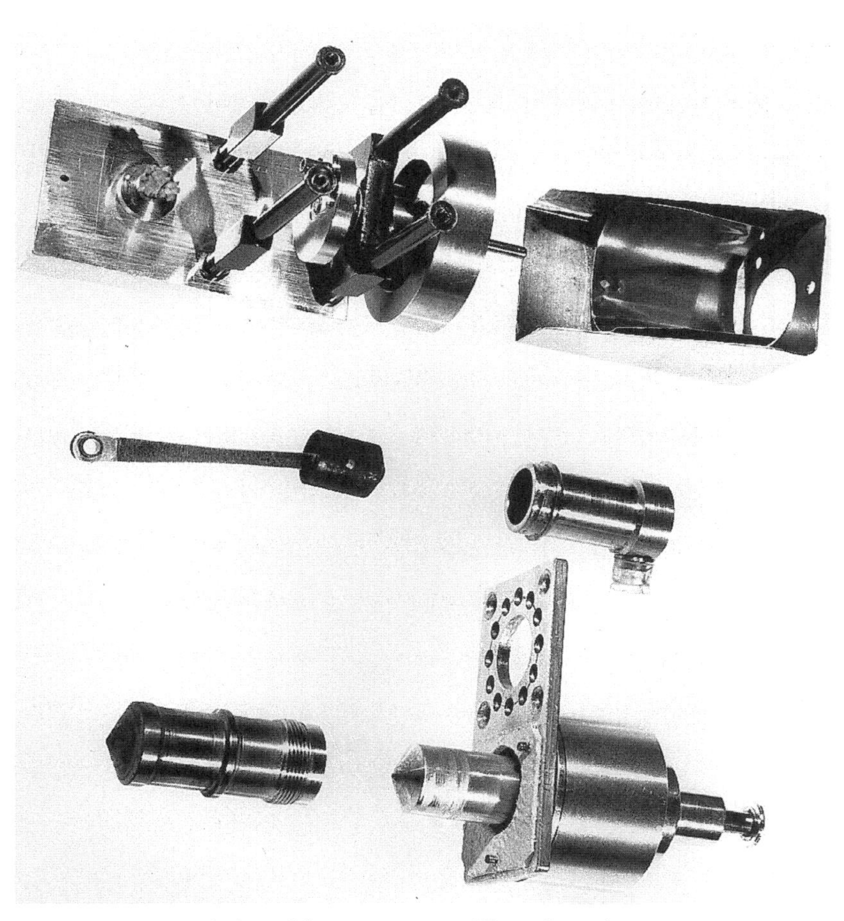

A view of the components of Fauvel's engine.

An article[8] describing Tapper was published in 1982 from which many other Tappers have been made. Two noteworthy examples are shown on the following page.

[8] "Tapper-a hybrid Stirling engine", Senft, *Engineering in Miniature*, 1982, p.101-102, 130-131, 173-174.

This fine running Tapper was built by the late Mick Fox.

This Tapper was constructed by J. J. Wood and displayed at the 1986 Model Engineer Exhibition. The judges reported that the engine "...proved to be very popular. It was run many times and just for fun we connected it to the brake. At 0.11 watts it should easily drive a small boat and with its tiny spirit flame it was possibly more efficient than any of the competition engines!"

L-27 is a Ringbom engine designed for operation at low temperature differentials. This engine will run for several hours on the thermal energy contained in two cups of hot water poured into its insulated base.

Chapter 3. L-27

One of the most fascinating types of Ringbom is the low temperature differential (LTD) variety. An LTD engine can be loosely defined as one that can operate when the difference in temperatures between the warmest and coolest sections of the engine is less than 100° C (or 180°F). The components of an LTD Ringbom and its basic sequence of operation are the same as for an engine like Tapper, but it is built to a radically different geometry. Its displacer is large in diameter and short in stroke with many times the swept volume of the power piston. This kind of geometry is precisely what makes an engine capable of running at small temperature differences between its hot and its cold ends. A low compression ratio is the secret to getting a Stirling engine (or any engine) to operate at low temperature differentials. The lower the compression ratio, the smaller the temperature difference at which the engine is capable of running.[1]

That explains the shape of the engine described in this chapter. Nicknamed the "L-27" [2], this little Ringbom can operate for hours from the thermal energy in a few cups of hot water, or it can be setup to run on passive solar energy. With near boiling hot water as the heat source, the engine operates at over 700rpm. In the sun, with a temperature difference of about 60°C (108°F), the engine has run at over 500 rpm and delivered a peak power of 1/4 Watt at 270rpm.[3]

---

[1] For more information on the entire field of LTD engines cf. *An Introduction to Low Temperature Differential Stirling Engines*, Senft, Moriya Press, 1996.

[2] The "L" comes from the Lindbergh Foundation, the sponsor of a research project under which the prototype of the engine was designed and built. "27" was the page number in the logbook for the project on which the design was completed.

[3] Test results and technical aspects of the L-27 are detailed in the book *Ringbom Stirling Engines* , Senft, Oxford University Press, 1993.

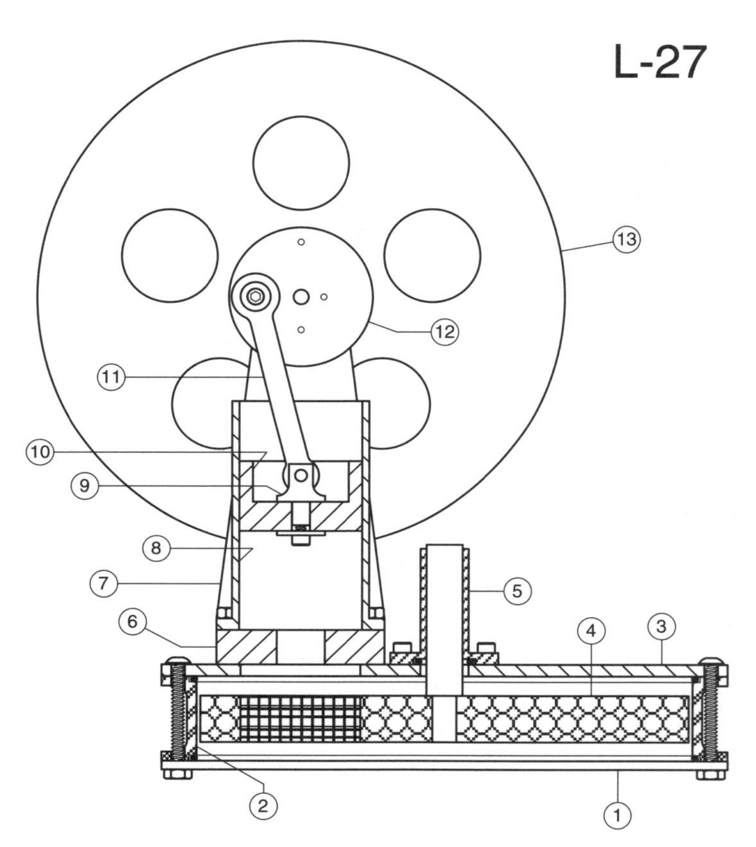

L-27 Parts List	
Item No.	Description
1	HOT PLATE
2	DISPLACER CHAMBER RING
3	COLD PLATE
4	DISPLACER
5	DISPLACER DRIVE CYLINDER
6	CYLINDER BASE
7	BEARING PLATE
8	CYLINDER

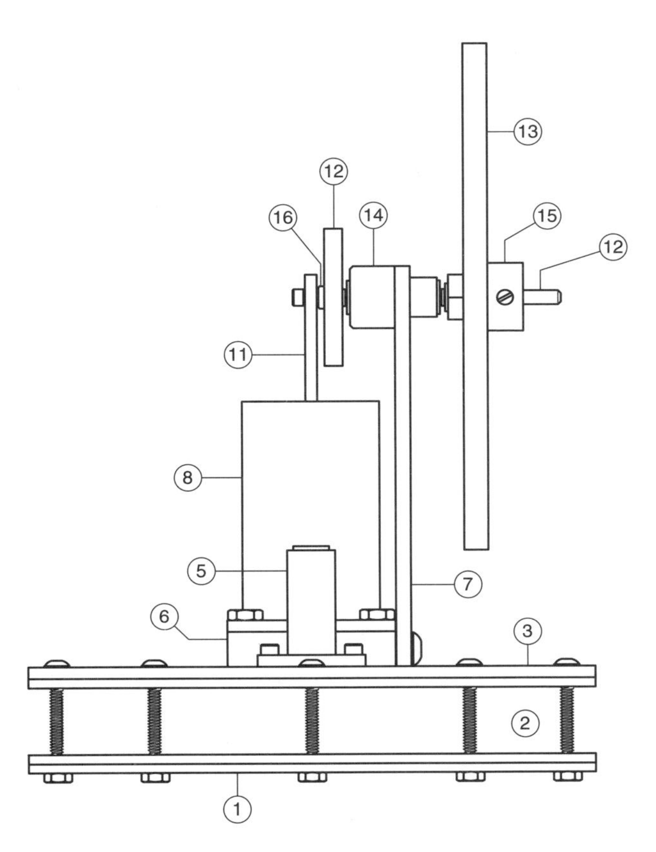

L-27 Parts List	
Item No.	Description
9	PISTON YOKE
10	PISTON
11	CONNECTING ROD
12	CRANKSHAFT
13	FLYWHEEL
14	BEARING HOUSING
15	FLYWHEEL HUB
16	CRANKPIN

The L-27 is a well-tested design. Many copies have been made since its general description was first presented in technical papers and reports.[4] The constructional methods used are similar to those for Tapper and the author's "N-92" LTD engine.[5] It does demand more than the usual amount of care and patience that goes into making a conventional model Stirling engine, first because it is a Ringbom, and second because it is a LTD engine. Friction and leakage must be absolutely minimal in this kind of engine. It is recommended that one have made a successful Stirling or two, and preferably a Ringbom like Tapper, before taking on this project. Properly built, the L-27 is very robust and a dependable performer.

Displacer Chamber Ring

The displacer chamber body is machined from solid acrylic. Clear plastic of course is the best choice so that you will be able to watch the lively displacer motion. A 6" square of 1" thick material is required. First machine out the center hole. Mount the material to a faceplate with four bolts in the corners and space it about 1/8" away from the faceplate to avoid cutting into it. You can save time and material for a future project if you begin by cutting out as big a disc as you can with a trepanning tool. Then bore the hole to finished size. Face around the hole to bring the thickness there to about .911". This is the dimension to which the original turned out, but is not critical; ten or fifteen thousandths either way will be fine. The o-ring grooves can be machined now or later. The first groove is easy at this point. To do the second one, the part will have to be reversed on the faceplate and recentered using a dial indicator and a good measure of patience. You can wait until later after a mandrel has been made which is need-

---

[4] Cf. e.g. "A solar Ringbom Stirling engine", Senft, *Proc. 21st Intersociety Energy Conversion Engineering Conference*, Paper 869112, American Chemical Society, 1986, or "A direct solar Stirling engine", Senft, *Journal of the Washington Academy of Sciences*, 1987, Vol. 77, p.183-189.

[5] Op. cit., *An Introduction to Low Temperature Differential Stirling Engines*, Part III.

ed anyway for machining the outer profile of the ring.

The circle for the bolts can be drilled next. If you have a carriage drilling spindle and means for indexing the lathe spindle, these holes can be drilled before removing the part from the faceplate. Another sure method is jig drilling on the milling machine by calculated coordinates or a rotary table. Laying out, centerpunching, and drilling on the drill press is also possible but must be done carefully to ensure the displacer will be centered. Alternately, the cold plate can be made first, centered over the ring bore, and the holes spotted through from the plate. The centering can be accomplished by turning a large washer to fit the ring bore with a 1/2" hole in the center to match the centerhole in the plate; a short 1/2" rod or dowel pin pushed through the two plates will align the centers.

The machining of the outside of the ring is best carried out mounted on a wooden mandrel as the ring is too thin to be held in a 3 or 4-jaw chuck. First rough cut the outside to about 6" in diameter on a band or jig saw. Secure a rough sawn wooden disc to the lathe face plate and turn a shoulder on it so the ring is a snug push fit on it. The shoulder should only be about 7/8" long so the face of the ring stands out slightly from it. The ring can be clamped secure from slipping by driving three or four wood screws into the end of the mandrel with fender washers or wooden strips to bear on the end of the plastic ring. The OD can be then turned to size and the recess machined to the finished dimensions. The recess dimensions are not critical; the purpose of the recess is to thin the wall to minimize heat conduction through it from the hot to the cold end. To cut the o-ring steps at this point, the ring can be clamped in the same way, by screws and washers, but bearing on the flange of the ring next to the faceplate. The angle of the o-ring step is not critical here; it is there to keep the ring from slipping into the chamber. What is important is the finish of the groove surface. It must be smooth to ensure good sealing and the ring must stand at least .005" proud of the ring so the plates can squeeze it down to form an adequate seal.

Finish the ring by wet sanding and buffing the inner and outer surfaces to bring back their full transparency; alternately, a liquid or paste plastic polish can be used. Be careful in polishing not to touch the o-ring grooves so as not to damage their geometric truth and so impair their sealing ability.

If by chance some 1/8" wall plastic tubing with a 5-1/4" ID is available, a displacer chamber can be made from this. It will be without the flange of course, along the lines of the "N-92" engine.[6] This approach is not as positive as the flanged ring, but it works if reasonably accurate tubing can be had. The bolt circle in the plates will have to be changed to suit the tubing; they should just clear the OD of the ring and thus center it to the plates. Don't worry that the ID is a little larger than the 5.17" called for in the drawing. Just make the displacer correspondingly larger so that it has the about same radial clearance to the ring. The engine will work fine because the little extra diameter will not change the compression ratio appreciably . You will still have to make a mandrel to mount the tubing for facing the ends and cutting the o-ring steps; if there is not enough friction in the fit to prevent the ring from slipping during machining, tape the ring to the faceplate or put a band clamp around the ring.

Cold and Hot Plates

The engine plates are made from aluminum. Choose alloy 6061 if you will want to have them anodized for protection and color. Untreated plates in contact with water will sometimes corrode which spoils their appearance, but if that does happen, they can be made to look good again, maybe better than new even, with an epoxy base paint. The inside surfaces are no problem.

Aluminum is a good material for this application. The plates have only to be thick enough to resist bulging. The pressure change generated in L-27 is very low, but the area of the plates is large so there is a tendency for the plates to flex, and this has an adverse affect on

---

[6] *Ibid.*

the pressure-volume cycle. A thickness in the present case of 3/32" is adequate, which is what is called for on the lower or hot plate. The cold plate is specified to be 1/8" thick so that there is a little extra material to take countersinks and threads.

It is important that the plates be smooth and free of nicks and scratches particularly where they contact o-rings. They also need to be flat, at least enough so that when the screws are tightened, the plates will pull down against the o-rings to seal all the way around.

The cold plate carries many holes. Their position is critical only insofar as mating parts will fit. It is important to ensure that the 1/2" hole will be centered fairly accurately over the bore of the chamber ring. This is so the displacer will not rub against the chamber ring. The displacer must be absolutely free.

Although not shown in the drawing, you may wish to put one more hole in the cold plate to serve as a compression release vent. It is not essential on this engine, but can be helpful. Normally it is plugged by a screw with a gasket or o-ring under its head as on the N-92 engine.[7] It can be removed to vent the engine interior to the atmosphere during assembly to check that the movement of the piston and its drive mechanism are free. It also can be an aid when starting up by enabling excess air to be released. For this engine, a 10-32 tapped hole will do fine. A convenient location is diametrically opposite the cylinder.

The holes in the hot plate only need to line up with those in the ring. They can be spotted through the cold plate. On the original engine, nuts were used on the hot plate side. It is more convenient in using the engine however to have an entirely flat hot surface. For this, the hot plate holes can be tapped 6-32 to take the assembly screws which are trimmed to the exactly length required. In this case it is better to make the hot plate 1/8" thick also.

---

[7] *Ibid.*, p. 51 & 56.

Displacer Drive Cylinder

The gland or cylinder in which the displacer rod works is shown in the drawing as being machined from solid brass or aluminum. You may be able to find some tubing with an accurate smooth 3/8" dia. bore in which case the flange can be made up separately and soldered on. A little bit over or under the 3/8" bore can be tolerated; plus or minus .010" wouldn't make a big difference in performance. On the original L-27, a commercially made glass cylinder[8] was cemented to a flange machined from aluminum. In any case the bore in which the rod operates must be true and smooth. If machined from solid, the bore should be finished by polishing, honing, or lapping as discussed in the previous chapter for the cylinder of Tapper.

The recess in the base of the flange accepts an o-ring for sealing. Pick one out at your local hardware store having about a 1/2" ID and then counterbore the flange to accept its OD. Leave ample space around it, but not so much that the o-ring can go off center enough during assembly to rub on the displacer rod. The counterbore can be generous in depth also; just five or seven thousandths of compression will adequately seal the o-ring. Remember that a smooth finish is necessary for the ring to seal.

Displacer and Rod

The displacer rod is turned from solid graphite rod to a close free fit in the displacer drive cylinder. On the prototype, instead of machining a rod, the piston that came with the glass dashpot cylinder was used. But this required an additional guide pin to keep the displacer centered. The solid rod is simpler, even if you also use a dashpot cylinder. In making the rod, follow the general advice given for machining the piston on Tapper in the previous chapter. The rod

---

[8] Manufactured for use as a dashpot by the Airpot Corporation, 35 Lois St., Norwalk CT 06851. Phone (203) 846-2021. FAX (203) 849-0539.

should be drilled out as shown in the drawing to lighten it. The displacer end is stepped and grooved to secure it to the displacer body by epoxy.

The displacer body is made from 1/2" thick Styrofoam. Ordinary white Styrofoam was used on the prototype with excellent results, but the blue variety has been used in other LTD engines and seems to be more stable. After rough cutting the disc, glue the displacer rod in place with epoxy. If the rod is not square to the displacer surface, the displacer will tend to wedge itself in place when it comes into contact with a plate at the end of its stroke; this could retard displacer motion and adversely affect performance. On most LTD Stirlings, I use a v-block resting on the displacer surface to fixture rods square while the glue is curing.

The outer diameter of the displacer can be finished by hot wire cutting or against a disc sander. Sanding is more controllable, but generates fine dust which is difficult to clean off. The dust must be removed before the engine is assembled lest it jam between the piston and cylinder and retard free motion. In finishing the OD to size, use the rod as a reference to get concentricity, but protect it by slipping the cylinder over it. Aim for a radial clearance of about .040" (1mm) between the displacer periphery and the chamber ring bore. If the gap is more, the engine will still run, but more air will flow around the displacer rather than through the regenerator elements, which will correspondingly reduce the speed and power of the engine. A smaller gap would be better, but if too small, there will be a greater chance of the displacer rubbing against the ring.

The engine will run without the regenerator elements, but they improve the performance by roughly double the speed and power . It is therefore worth the trouble of putting them in. If you are tempted to try the engine without them, you will need to perforate and groove the top of the displacer so that it does not block off the power cylinder hole when it is against the top plate; this would prevent the piston from moving out on the expansion stroke.

The holes for the regenerator discs can be cut with a sharpened piece of tubing pressed into the displacer and turned back and forth by hand. The regenerator elements can be cut the same way from air conditioner filter foam. Look for some at your hardware store that is 1/2" thick (or just under) and has fine strands and an open structure. Glue them in place making sure they do not protrude above the surface of the displacer body.[9]

Cylinder Base

If you make the cylinder base next, you will then be able to assemble the displacer section and check that everything made so far fits and works as it should. The 10-32 tapped holes on the side are for mounting the bearing plate. Only two are required, but four are shown. This is so that if you ever decide to change the orientation of the shaft and flywheel position after the engine is assembled, you can just remount the bearing plate in the other pair of holes, and will not have to remount the cylinder base on the cold plate which would require dissembling the entire displacer chamber unit. When making this part, be sure to mill or face the surfaces where the bearing plate will mount so that it will stand up square.

If you intend to use an Airpot dashpot for the engine power cylinder as on the prototype, then you will want to make the center hole larger than 1/2" to accept the standard nose of the Airpot. The four 6-32 holes are dimensioned on the drawing to accept the standard Airpot flange.

The base mounts to the cold plate with four 6-32 flat head screws. The screws should be just about 7/8" long to protrude from the top of the block enough to accept the cylinder flange, which is 1/8" thick, and the nuts to hold it. The block should be sealed to the cold plate using a gasket or a sealant. Alternately, an o-ring groove could be machined in the block. In any case, use a sealant (e.g. Loctite blue

---

[9] For more information on LTD displacers, cf. *An Introduction to Low Temperature Differential Stirling Engines*, op. cit..

or epoxy) in the countersinks so that leakage cannot occur under the heads of the screws . When assembling, orient the block so that the bearing plate holes will not be next to the displacer rod gland.

Displacer Section Assembly

The large o-rings for sealing the plates to the ring are made from o-ring cord stock (or from larger o-rings). Mill, bearing and hydraulics suppliers stock this material. Use nominal 1/16" stock (actual size is about .070") in as soft a durometer as available. I used the red silicone type. Cut the ends square by feeding the material through a .070" diameter hole drilled squarely through a piece of metal or plastic plate and sliding a sharp razor blade across. When trimmed to the correct length, glue the ends together with instant glue. I use a Teflon block with a v-groove milled in as a guide to align the ends when gluing up o-rings.

Mount the displacer drive cylinder to the cold plate with 4-40 screws trimmed in length so that they will not protrude and hit the displacer. Don't forget the o-ring. Put a little sealant on the screw ends if they are not tight to reduce air leakage past the threads.

Now place one of the o-rings into its groove on the chamber ring and place the cold plate down on it. It helps to have a few screws placed in the ring from the underside to guide the plate to the correct position. Too much sideways movement when assembling will dislodge the o-ring from its step and you have to start all over. It also helps to have made the o-rings on the large side so that they will not so easily move inward. A very light coat of silicone grease can be smeared on the o-rings to make disassembly easier in the future if ever required.

Lightly clamp or tape the ring to the cold plate to keep everything aligned during the rest of the assembly. You can now remove the alignment screws and turn the assembly over for easier working. Slip the displacer rod into its drive cylinder and check that the displacer approaches the plate closely and does not bind. Put the other o-ring in place in its groove and lower the hot plate down. Now you can put in the ten 6-32 screws to finish the assembly of the displacer unit. If you have access to Nylon screws, they will cut down on heat conduction from the hot to the cold plate, but if not, stainless steel will be OK as was used on the prototype. They do not have to be tightened excessively, only just enough to seal the o-rings.

It is a good idea to check the displacer for proper action at this point. You will have to rig up a temporary tubing connection to the cylinder base hole for this. With just a tiny breath of pressure, the displacer should pop up against the cold plate. Likewise if the unit is turned over, just a little suction should bring the displacer up against the hot plate. Action should be extremely free and lively. If it is sluggish or sticky, the engine will not work as it should, or maybe not at all.

Cylinder and Piston

The fit between the piston and cylinder must be exceptionally close and free. Leakage and friction must be absolutely minimal. The glass cylinder used on the prototype is a standard commercial item,[10] but the drawing gives dimensions for a fabricated version. Brass, bronze, stainless steel, or aluminum are suitable materials. Steel is not recommended because the piston must run dry to avoid viscous friction, and then rusting would be a problem. As with Tapper, the cylinder bore must be true and smooth. It should be finished by honing or lapping. The cylinder on this engine is large enough so that your local automotive shop might be able to hone it for you. The wall thickness can well be increased over that specified in the drawing for greater geometric stability. Likewise, the flange could be made heavier too, say 5/32" thick. As with the displacer drive cylinder, if suitably round and smooth tubing is available, a separate flange can be made and fitted.

---

[10] Also manufactured by the Airpot Corporation.

It is not essential to adhere to the exact bore specified, since the stroke can be adjusted to compensate. The swept volume is the important parameter here rather than the bore and stroke combination that produces it. Actually, the swept volume is adjustable on the engine to adapt it to run optimally on whatever temperature heat source being used. At high temperature differentials, as when the engine is placed over boiling hot water , a piston swept volume of 25cc will give the most power. After the engine has run for a while and the water has reached the lukewarm stage, you will have to decrease the stroke of the engine to a swept volume of 15cc or 10cc to keep the engine running. As long as the power cylinder can give swept volumes in this range, from about 1.5 cu.in. to 0.6 cu.in., the particular bore and stroke combination you use is not important.

Again, graphite is the best material to use for the piston on LTD engines. Follow the recommendations given above for Tapper. The dimensions shown in the drawing are roughly those of the standard commercial dashpot piston.[11] If you are making your own piston, it can with advantage be made longer to resist wear and leakage.

Piston Yoke

The piston yoke is machined from aluminum for light weight. Begin by cross drilling and reaming 1/8" diameter through the end of a length of 1/2" diameter bar stock. Mill the slot next. Cut off to near finished length and chuck in the lathe for machining the spigot. There is not a lot to grab onto for this chucking, so it helps to make a split bushing into which the part can be inserted. Turn the spigot to an easy fit in the hole in the piston center. Finish by filing away unneeded material as shown in the drawing.

---

[11] If you use a standard Airpot unit, the factory made rod will have to carefully removed by drilling out the riveted end and replaced by the yoke and connecting rod specified in the drawings. Airpots can be special ordered with no rods in the piston.

Bearing Plate and Bearing Housing

The bearing plate is cut from 5/32" thick aluminum plate. It is drilled at the bottom to mount to the cylinder base with two 10-32 machine screws. At the top it carries the crankshaft main bearing housing which is turned from round bar. It is made a press fit into the bearing plate, or can be secured by retaining compound or epoxy. The bearing housing is shown as reamed through 5/16" diameter which will accommodate flanged style ball bearings with a 5/32" bore. Unflanged bearings can also be used of course, but will require a counterbore or a sleeve pressed in. Use shielded, but not sealed, bearings . The shielding does not absolutely seal the bearing but is very effective at keeping dust out and avoids the friction of seals. If you have open unshielded bearings to work with, you can machine washers to fit over the shaft and large enough to cover most of the open part of the bearing. Machine a shallow step on the washer face for clearance so it cannot contact and rub on the cage or outer race. If the bearings come greased, wash them out in solvent and lubricate lightly with thin watch or clock oil.

Crankshaft, Crankpin, and Connecting Rod

The crankshaft is made by press fitting a length of 5/32" diameter rod into an aluminum disc. For added security, I pinned the two together also. The easiest way is to drill in from the end 1/16" diameter half in the shaft and half in the disc; drill about 5/32" deep and push in a short pin flush with the surface of the disc. The disc is drilled through and tapped 4-40 in four locations to take the crankpin screw. The dimensions shown will give piston swept volumes of 25cc, 20cc, 15cc, and 10cc.

The crankpin is the same style as on Tapper. It is a simple short bushing drilled to take a 4-40 screw which secures it to the crankdisc. The large end of the bushing is held against the crankdisc . The small diameter is turned to a loose fit into a 5/32" bore ball bearing which is carried in the large end of the connecting rod. The chamfer on the crankpin should be cut so that shoulder clears the outer race and shield of the bearing. The screw used on the prototype is a socket head screw also chamfered for clearance. It was fitted with a knurled knob so that it can be removed and replaced without tools.

With the bearing plate and crankshaft in place, and the piston and its yoke made, double check the length of the connecting rod needed for your engine. At the largest stroke, the piston should easily clear the cylinder base. A clearance of 1/16" will be fine.[12] Both ends of the connecting rod are fitted with ball bearings to reduce friction as much as possible. The little end bearing has a 1/8" bore. If the OD of your bearings are larger, you will have to contour the rod outline to suit of course.

When assembling the little end of the connecting rod in the yoke, spacers will be required on each side to center it and prevent it from rubbing on the sides of the yoke slot. This slot is 1/4" wide, so make the spacers a nominal 1/16" wide but about .005" narrower to allow for any little misalignment. The spacers can most conveniently be parted off from standard .015" wall brass tubing. If none is at hand, turn some brass rod to 5/32" diameter, drill .128", and then part off a couple of washers with a thin cutoff bit. The top slide can be set parallel to the lathe bed and its micrometer feed can be used to get the width just right. The tool should be very sharp so as not to leave a burr which will be difficult to remove. The wrist pin is a length of 1/8" diameter drill rod; polish it down to a very easy fit through the bearing bore.

Flywheel and Hub

Turn the flywheel hub from aluminum or brass bar stock. The drawing shows a thread so that a nut can be used to secure it to the fly-

---

[12] There is no need to be overly concerned with a little extra dead space in LTD engines because there is so much volume in the displacer section.

wheel disc, but this can be omitted and the hub glued to the disc. The five holes in the disc are for appearance only and can be omitted. The hub is secured to the crankshaft by an 8-32 setscrew. Turn a short brass cylinder to drop into the setscrew hole to act as a pad against the shaft. This will keep the shaft free of burrs so that it can be removed easily. Short spacers will be necessary against each main bearing to avoid parts from rubbing where they are not supposed to touch; make them just as those on the wrist pin.

Final Assembly

Mount the cylinder on the assembled displacer unit with 6-32 nuts on the screws protruding from the base block. Seal the cylinder to its base with a soft gasket or sealant.

Next attach the connecting rod and yoke assembly to the piston. Use a thin gasket under the yoke to seal and cushion it against the piston; electrical tape is a good material to use here. Punch a 3/16" hole in a piece of tape, stick it onto the yoke, and trim it around. The piston washer should be the same diameter as the yoke base to distribute the load on the piston end uniformly. Cushion the washer face against the piston also with tape or paper . Tighten the 4-40 screw firmly so that the yoke seals against the piston, but avoid excessive torque so as not to unnecessarily stress the graphite piston.

Now you can insert the piston into its cylinder . Here is where the compression release vent comes in handy; without the vent, you will have to gently push the piston in until enough air has leaked out to let it stay inside the cylinder. The engine should be so free of leakage that this should take a half of a minute or so.

Check the action of the displacer at this point. With the piston at about midstroke in its cylinder (and the vent plugged of course) just a little push inward on the connecting rod should be enough to move the displacer upward. On the prototype, a move inward of just 1/8" will lift the displacer all the way up. When the piston is quickly jiggled up and down a 1/4" or so, the displacer should rapidly pop up and down. This is the lively action the engine must have to run successfully.

It is instructive at this juncture to experience the action of the displacer with a little heat applied. Set the engine down over a pan of hot tap water and move the connecting rod up and down by hand. As the displacer responds, you will feel the pressure rise and fall on the piston, and be able to sense the processes described in chapter one taking place.

With everything working as it should, all you need to do is hook up the connecting rod to the crankshaft. Open the compression release and check that the motion of the piston is absolutely free of any binding. This is best done without the flywheel mounted since its inertia can mask small tight spots.

To run the engine, you will need a reservoir which will hold a few cups of hot water and on which the engine will rest stably. For the first test, you will probably just hunt something up in the kitchen that will do, but eventually you will want a good solid well insulated base. I made one from Styrofoam. In any event, make sure the engine rests securely in place. Heat up some water, pour it into your reservoir, set the piston stroke to the 25 or 20cc location, and wait a minute for the bottom engine plate to warm up.

Give the flywheel a little turn to initiate displacer motion. If the engine plate is not warm enough yet, it may just oscillate back and forth. The oscillations will get larger and larger as it warms up and eventually it will go over center and start running in one direction. As with Tapper it may once in a while get stuck on center, but then it is ready to go and a nudge off center will set it to running.

If the water is very hot, it may happen that the first time the displacer pops up, the piston will move up and stay up, locked in position by high pressure in the engine. This happens because there is too much air inside the engine, and you can open the vent to let some out, or just patiently wait for enough to leak out. This also happens if the piston stroke is set too small for the temperature of the source.

As the water cools down, the engine will run slower and slower.

Eventually it will reach the point where it cannot go over center anymore and it lapses into back and forth oscillation. The compression ratio has become too high for the diminished plate temperature differential. Reducing the stroke will allow the engine to operate again. When started on two cups of near boiling water in a well-insulated base, the L-27 prototype runs for about an hour . After reducing the stroke, it will go for another hour.

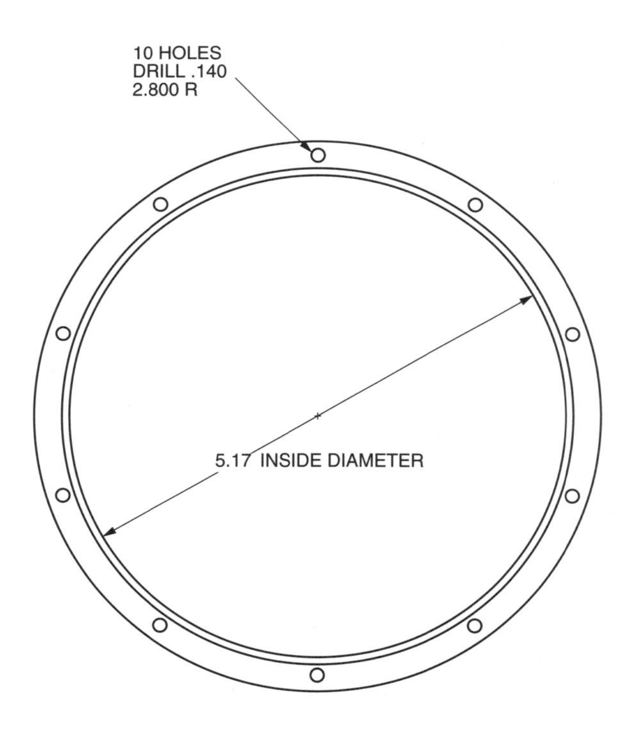

DISPLACER CHAMBER RING
ACRYLIC

Basic Specifications of L-27

Power piston swept volume: 0.6 in^3 – 1.5 in^3 (10 cc - 25 cc)

Displacer diameter: 5-3/32"

Displacer stroke: 13/32"

Displacer rod diameter : 3/8"

Displacer annular gap: .040" (1 mm)

Displacer mass: 6 grams

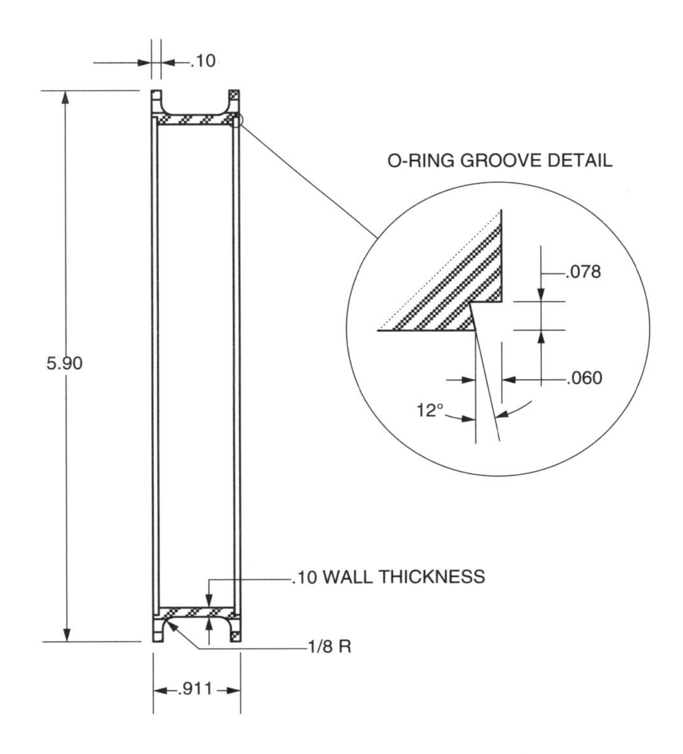

DISPLACER CHAMBER RING
SIDE VIEW

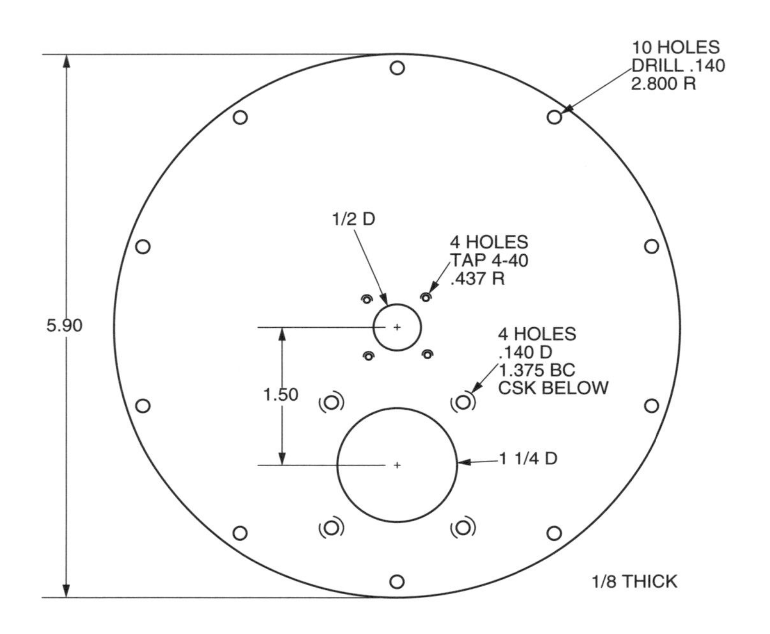

COLD PLATE
6061 ALUMINUM

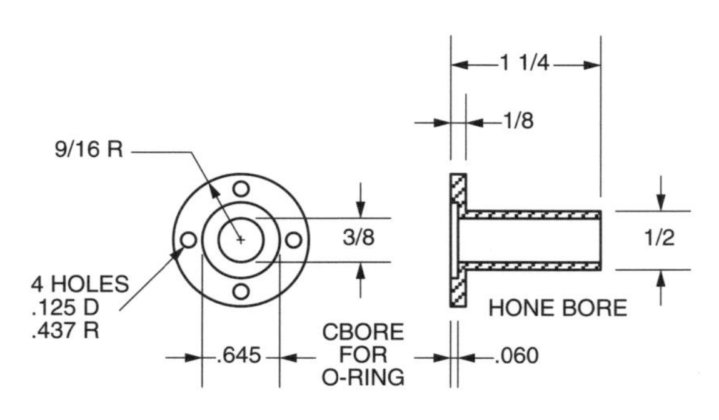

DISPLACER DRIVE CYLINDER
BRASS OR ALUMINUM

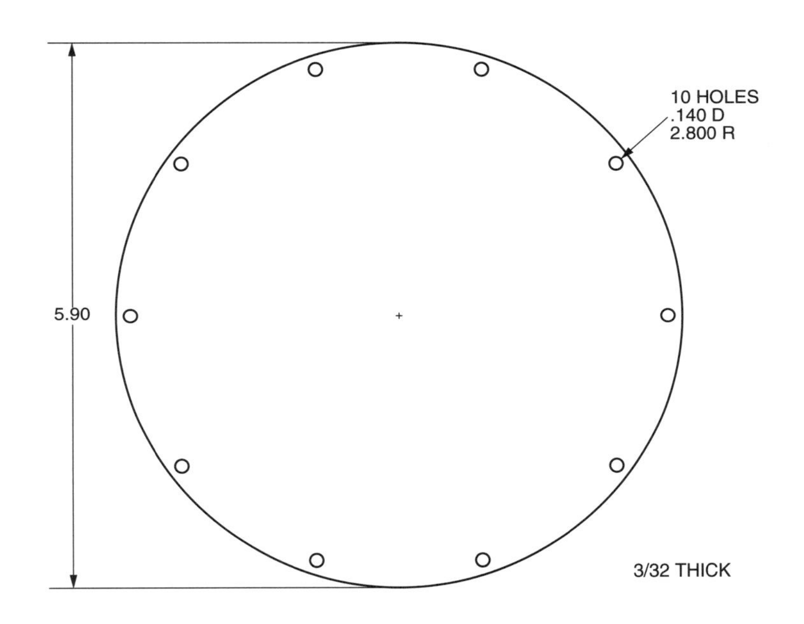

HOT PLATE
6061 ALUMINUM

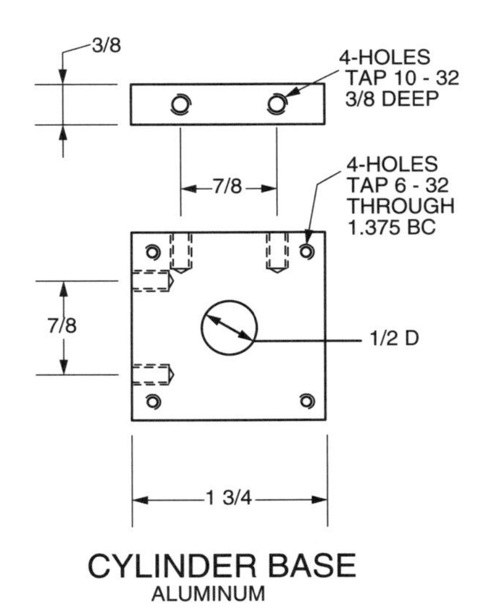

CYLINDER BASE
ALUMINUM

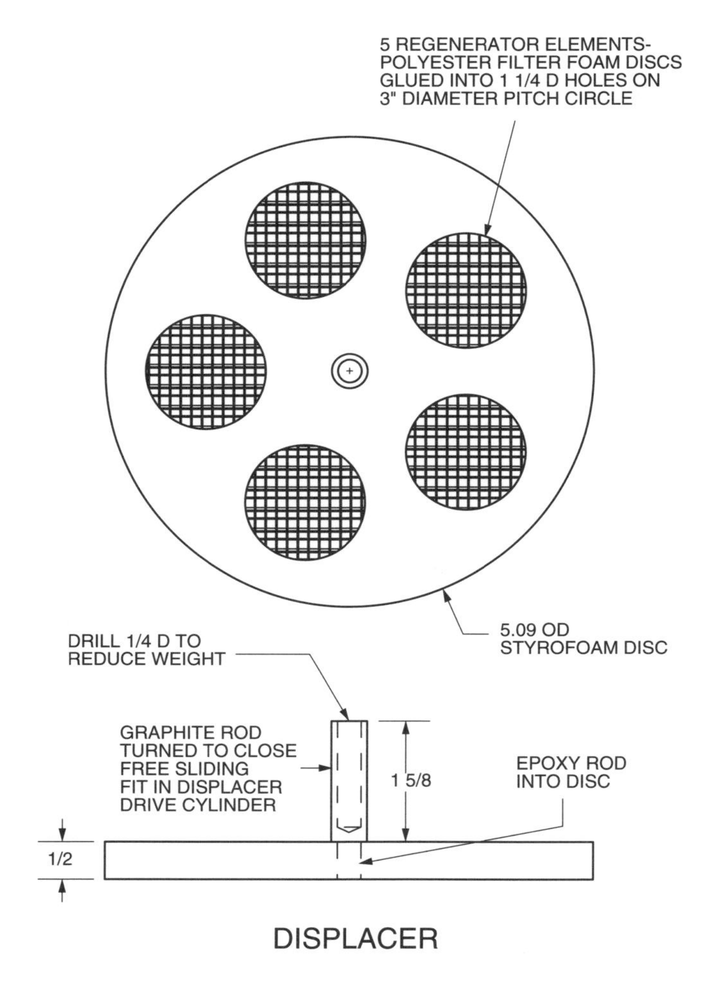
5 REGENERATOR ELEMENTS-
POLYESTER FILTER FOAM DISCS
GLUED INTO 1 1/4 D HOLES ON
3" DIAMETER PITCH CIRCLE
5.09 OD
STYROFOAM DISC
DRILL 1/4 D TO
REDUCE WEIGHT
GRAPHITE ROD
TURNED TO CLOSE
FREE SLIDING
FIT IN DISPLACER
DRIVE CYLINDER
1 5/8
EPOXY ROD
INTO DISC
1/2

DISPLACER

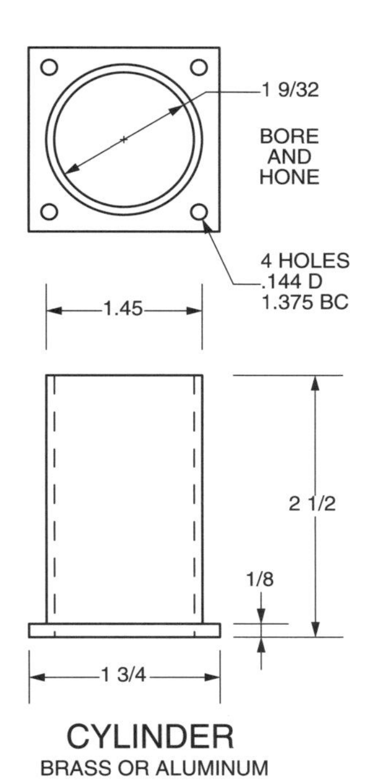

CYLINDER
BRASS OR ALUMINUM

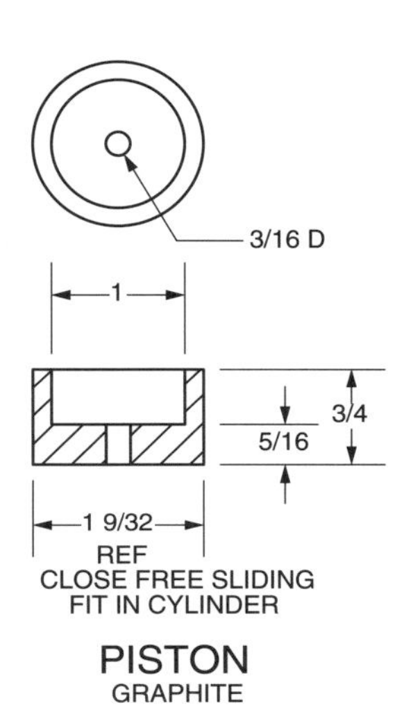

PISTON
GRAPHITE

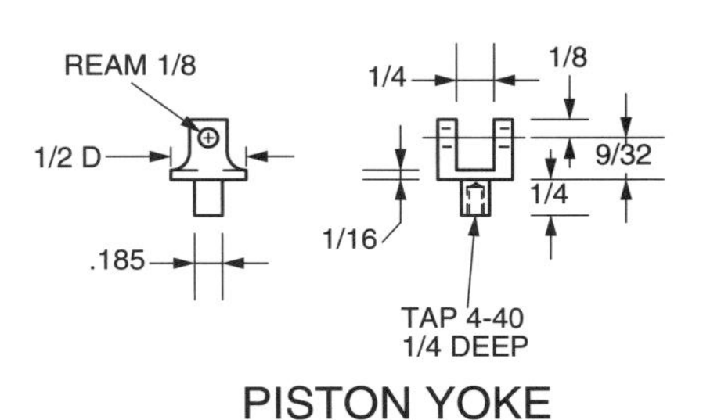

PISTON YOKE
ALUMINUM

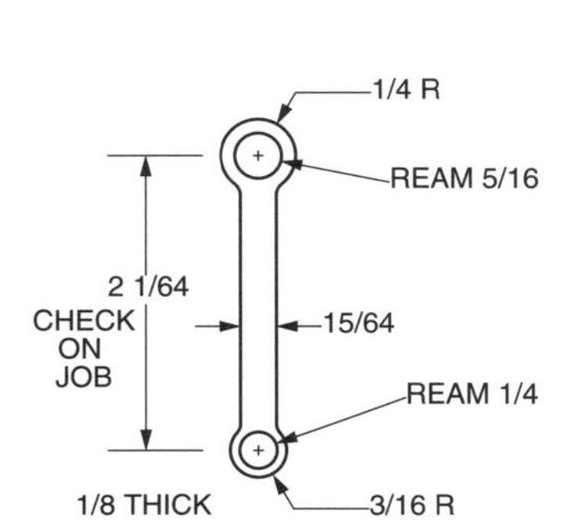

CONNECTING ROD
ALUMINUM

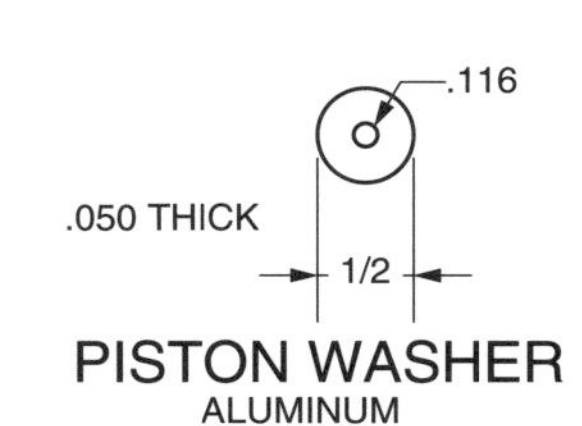

PISTON WASHER
ALUMINUM

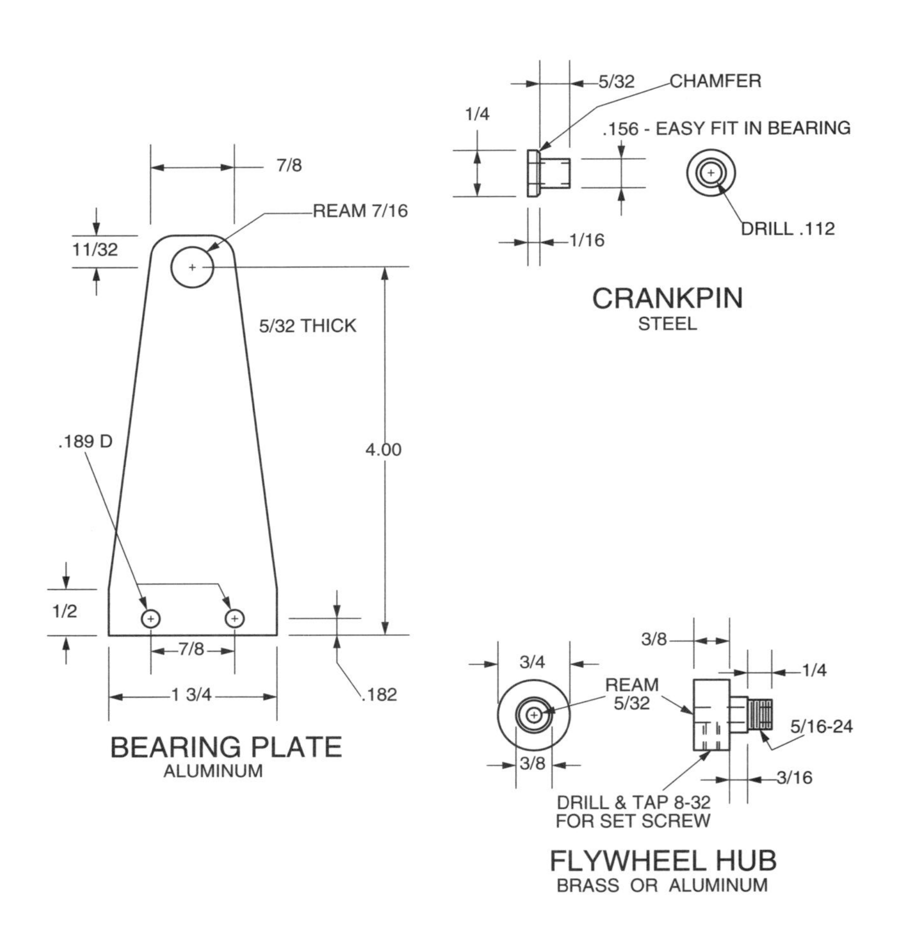
7/8
REAM 7/16
11/32
5/32 THICK
.189 D
4.00
1/2
7/8
1 3/4
.182
BEARING PLATE
ALUMINUM
5/32
CHAMFER
1/4
.156 - EASY FIT IN BEARING
DRILL .112
1/16
CRANKPIN
STEEL
3/8
3/4
1/4
REAM
5/32
5/16-24
3/8
3/16
DRILL & TAP 8-32
FOR SET SCREW
FLYWHEEL HUB
BRASS OR ALUMINUM

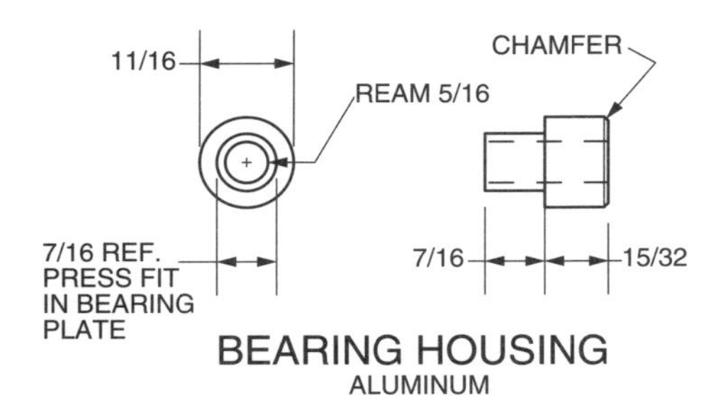
11/16
CHAMFER
REAM 5/16
7/16 REF.
PRESS FIT
IN BEARING
PLATE
7/16
15/32
BEARING HOUSING
ALUMINUM

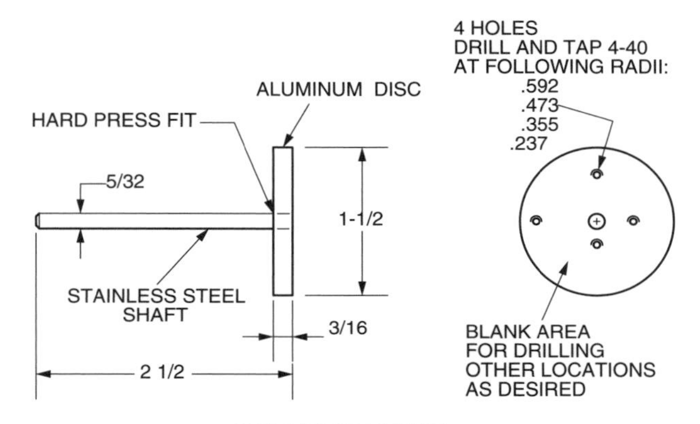

CRANKSHAFT

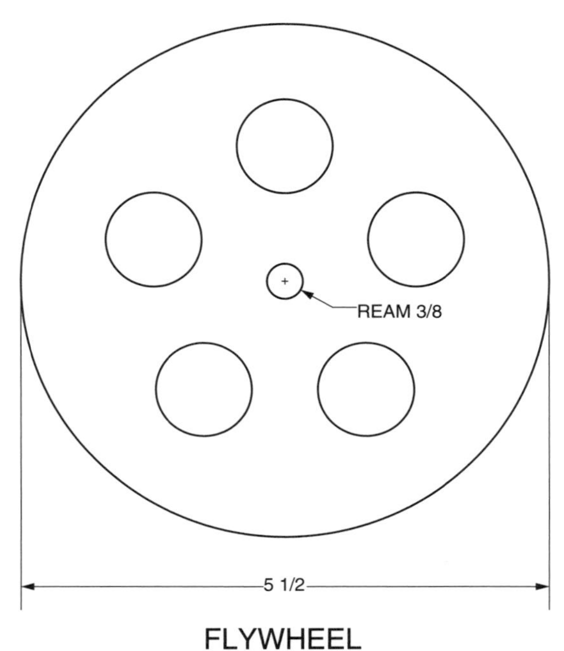

FLYWHEEL

1/4 THICK ACRYLIC

Some of the jigs and fixtures used in the making of the L-27 engine. In the center foreground is the wooden mandrel used to machine the displacer chamber ring.

The L-27 will run on solar energy by just blackening the hot plate and putting it in the sun. Shown here is the engine further modified for improved solar operation. The cold end plate was replaced with a square plate having deep widely spaced fins, and the hot end was equipped with a conical reflector to gather and direct more solar energy into the engine.

Another view of the L-27 engine. The two “ears” attached to the cold plate are brackets for mounting the engine for testing. They are also very convenient handles when picking up and moving the engine.

This very interesting fine running LTD Ringbom engine was made by Mike Shelley of Australia. After seeing the L-27, Shelley made this smaller version with a displacer diameter of just under 2". The displacer stroke is 11/32" and the piston swept volume is about 1.6cc.

Mike Shelley's friend Trevor Dowling made this fine looking LTD Ringbom. Trevor lapped glass tubing to make the power cylinder which has a bore of 1". The displacer diameter is 3-1/8" with a stroke of 1/2". The displacer rod is 0.35" in diameter. The aluminum hot and cold plates are machined with a threaded recess to screw onto the plastic displacer chamber.

Thumper is a single-cylinder Ringbom engine capable of high speed stable overdriven mode operation.

Chapter 4. THUMPER

There are two main types of displacer Stirling engines. The engines discussed so far in this book are of the separate or split-cylinder type. The piston and displacer each have their own cylinders in which to reciprocate. This kind of Stirling engine, also referred to as the "gamma" type, is the easiest to make, but there is a better configuration as far as thermodynamics is concerned. This is the single-cylinder or so-called or "beta" arrangement.

In the beta Stirling engine, both the piston and displacer operate in a common cylinder. Their strokes are made to overlap so that the piston and displacer virtually touch each other at one point in the cycle. This makes the most effective use of the hot and cold spaces of the engine, giving the beta engine better thermal efficiency and higher specific power.[1] The single-cylinder form of the Ringbom has the same thermodynamic advantages, along with elegant mechanical simplicity of the free displacer drive.

The Beta Ringbom

The basic single-cylinder engine concept has the piston forming the cold end of the displacer chamber. In the conventional beta Stirling engine, the displacer rod passes through the piston. This can also be done in the Ringbom form, but is not necessary because no mechanical connection to the displacer is required in a Ringbom. In the single-cylinder Ringbom, construction can be greatly simplfied by inverting the conventional arrangement. The displacer rod can be rigidly attached to the piston crown and the gland built into the cold

---

[1] For a detailed explanation of the beta type Stirling, cf. *An Introduction to Stirling Engines*, Senft, Moriya Press, 1993, p. 46-58.

end of the displacer. This allows the piston to be linked to the crank-shaft by the usual sort of connecting rod .

In order to work at all, this form of the single-cylinder Ringbom requires that the pressure inside the displacer to be equal to the mean or average pressure of the engine cycle. This condition is most easily assured by making the displacer rod tubular with a passage through the piston. The interior of the displacer will then at all times be at atmospheric pressure, or at whatever buffer pressure might be established within an enclosed crankcase.

Now it should be clear that if the pressure inside the engine goes above atmospheric or buffer pressure, the displacer will be pushed toward the piston. If the pressure inside goes below, the displacer will be pushed away from the piston toward the hot end.

This gives the same displacer drive action relative to piston motion as in the gamma Ringbom. Pushing the piston inward drives the displacer outward into the cold end of the engine. Pulling the piston outward, draws the displacer inward into the hot end of the engine. This is the same way the displacer responds in the split-cylinder Ringbom. Therefore, the engine will follow out the same sequence of operation. The big difference in operation is that in the single-cylinder engine, the displacer can ride along with the piston during the outward expansion stroke, from which it derives its principal thermodynamic advantages.

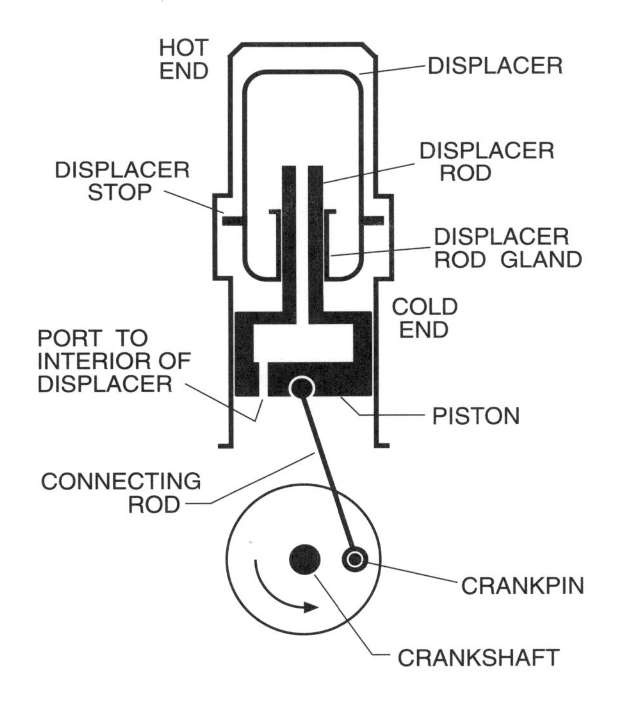

A schematic representation of the single-cylinder or beta type Ringbom engine. The displacer rod is attached to the piston and works in a gland carried by the displacer. The displacer stroke is limited by stops, one of which prevents the displacer from striking the hot end. The other stop limits the distance that the displacer can travel with the piston, which when properly located, improves the speed potential and stability of the engine.

Thumper

Thumper was the first single-cylinder Ringbom engine to acheive stable operation.[2] It operates in the overdriven mode up to 900rpm using a strong flame as the heat source. Like Tapper, it is quite entertaining while warming up. It will oscillate back and forth until it is good and warm, and then suddenly go over center and take off running steady, stable, and strong. Unlike Tapper, no part of the displacer is visible from the outside. All one can see moving is the crank, connecting rod, and piston. This engine is a real puzzle to someone seeing it for the first time!

Thumper was built as an experimental exercise, and is described here only for those wishing to take on an advanced project. It is more difficult to make than Tapper or the L-27, but most of the materials and techniques are the same as for those two engines. The displacer and its rod are considerably more complex as a look at the drawings will reveal.

These drawings are a record of the dimensions followed on the original. Because individual constructors of a engine project of this nature will no doubt wish to modify various details to suit their own equipment, materials, and preferences, only the methods followed in making the more unusual parts of the original will be detailed here. Hopefully this will be of assistance to those wishing to embark on their own adventures with single-cylinder Ringboms, as well as providing ideas and tips for engine building projects of all kinds.

---

[2] *Ringbom Stirling Engines,* Senft, Oxford University Press, 1993, p.138ff.

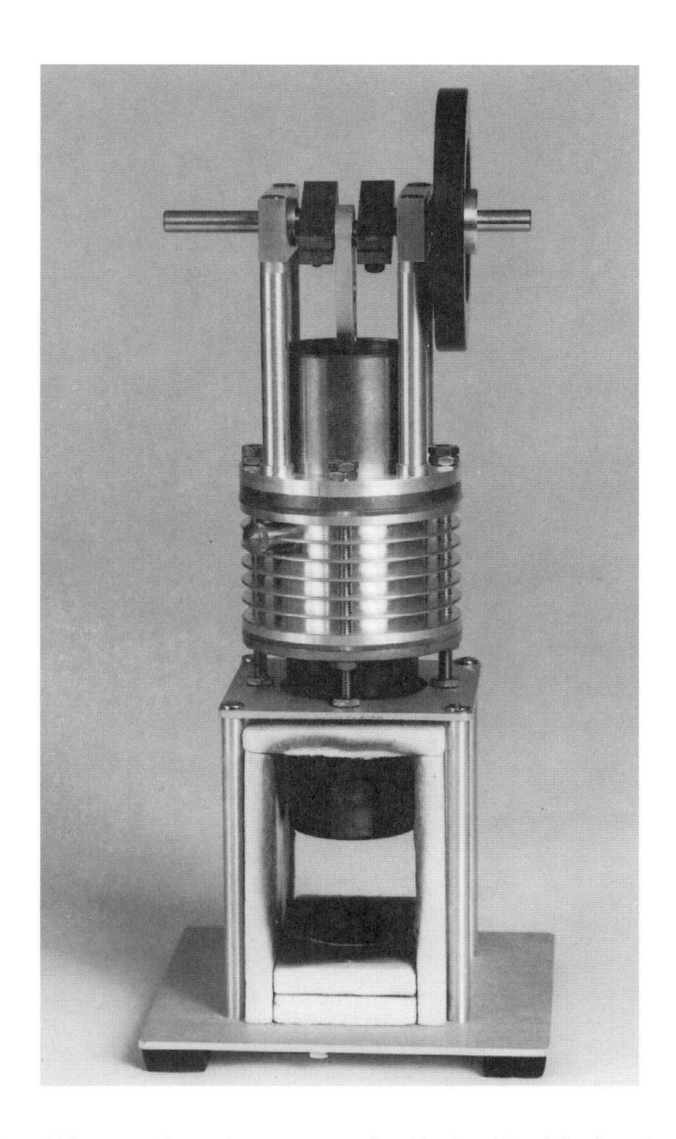

This view of Thumper shows the vent screw fitted in the side of the finned cold end. Removing this screw opens the engine space up to the atmosphere. This relieves compression so that the free movement of the crank mechanism can be checked. It can also be used to adjust the amount of air contained in the engine for faster starting.

THUMPER

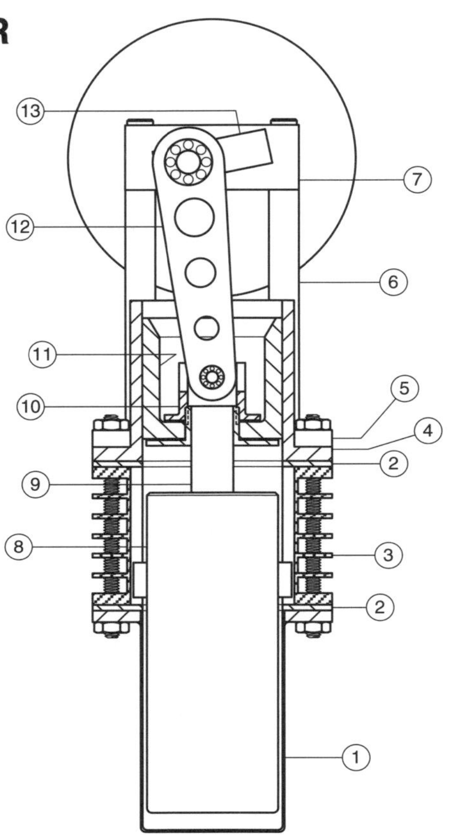

Thumper Parts List	
Item No.	Description
1	HOT END ASSEMBLY
2	GASKET
3	COLD END
4	CYLINDER
5	BEARING RING
6	BEARING PILLAR
7	BEARING BLOCK
8	DISPLACER ASSEMBLY
9	DISPLACER ROD

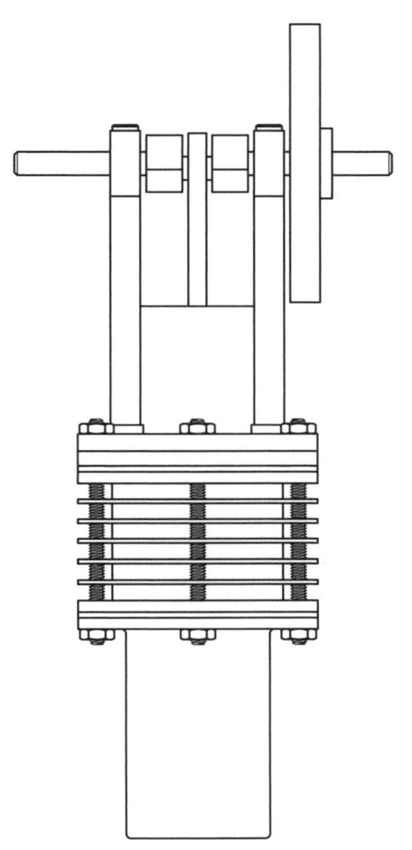

Thumper Parts List	
Item No.	Description
(10)	PISTON YOKE NUT
(11)	PISTON
(12)	CONNECTING ROD
(13)	CRANK SHAFT
(14)	DISPLACER SHELL
(15)	DISPLACER END
(16)	DISPLACER ROD BUSHING
(17)	DISPLACER BUSHING NUT
(18)	DISPLACER ROD CAP

Cylinder and Piston

A good place to begin construction of any engine is with its heart, the cylinder and piston. On the original Thumper, the cylinder was made from 660 bearing bronze. This is excellent cylinder material. It machines easily and wears extremely well. It is available from metal and bearing supply houses in the form of cored bars; in this case 2-1/2" OD by 1" ID stock was used. After machining was finished, the bore was lapped with an adjustable lap and non-embedding compound. A commercial barrel lap was used in this case, made by Acro. It was supplied as an 1-1/2" diameter lap, but the brass barrel was thick enough to allow it to be turned down to suit the 1.46" bore dimension decided upon for this engine. This odd size was chosen because that happened to be the inside diameter of a stainless steel drawn cup that was available for making the hot end.

The piston was made from a good grade of EDM graphite. The wall thickness of the piston is generous because it must support the displacer rod. The machining sequence followed for finishing the piston is given on the drawing.

Displacer Rod

The displacer does not contact the top of the piston, but rather the aluminum end of the displacer rod assembly. The rod was made first, honed true and smooth, and then the aluminum end made.

The best rod material is 416 stainless steel. It has a lower coefficient of thermal expansion than the more common stainless alloys so is less likely to grow too tight in the displacer bushing if the engine is run really hot. With other materials such as 304, mild steel, brass, or aluminum, the displacer bushing would have to be fitted looser when cold.

The outside of the rod had to be finished perfectly round, straight and smooth. Although lapping or finishing with a tool post grinder would work, a simple hone was made to do the job. As shown in the drawings, it is in the form of a pinch clamp with a small piece of oilstone epoxied in place. The stone and the two surfaces opposite it give three-points of support for the honing tool on the rod.

The hone is used in the same way as a lap. The work is run at a low speed, with plenty of thin cutting oil applied. The hone is tightened only enough to stabilize it on the rod, and run back and forth rapidly to produce a crosshatch pattern. A medium grit stone took out the tool marks quickly, and then the rod was lightly polished with 600 grit wet or dry paper. After the rod surface was finshed, the aluminum end was turned with a tight bore for the rod. There are no big stresses on the rod when the engine is running, so it is not likely to come apart but a pin could be put through the two pieces for added security.

Displacer Shell

On the original Thumper, the displacer shell was made from the body of a capacitor found at a local surplus center. Its outside diameter of 1-3/8" gives just about the right clearance to the hot end. A radial gap in the neighborhood of .040" gives enough flow area and insurance against the displacer rubbing the hot end due to any minor misalignments. The capacitor can was aluminum which is not the best material for displacers from a thermal standpoint as was discussed in Chapter 2. However, it makes for a lightweight displacer which is a good quality for a Ringbom engine.

Travel limits or stops were necessary to prevent the displacer from banging directly into the hot end. In this engine they took the form of ears soft soldered to the displacer shell. The ears are just bits of 3/16" diameter aluminum rod with flats filed as shown in the drawing. A good brand of aluminum solder was used for attaching them. The ears were drilled and pinned to the shell to hold them in place during the soldering. The displacer ears also limit the travel into the cold end. This arrests the outward movement of the displacer after it has travelled sufficiently far along with the piston; this provides more stable operation.[3] The ears make contact against 1/16" thick silicone rubber gaskets that seal the cylinder and hot end to the cold end.

Displacer End and Rod Bushing

The displacer end was a complex turning job. It was compounded by the requirement that it be light in weight, so its sections are thin. It of course caps off the displacer shell and carries a graphite bushing which rides on the displacer rod. This bushing is held in place by a thin slotted "nut" which captures it and presses it against an o-ring as shown in the displacer assembly drawing. Alternately, the bushing could have been simply cemented in place with epoxy as was done for Tapper, but this was an experimental project and modification or replacement would have been difficult if it was glued in the displacer.

The outline of the end piece was turned on the end of a length of 1-3/8" diameter stock, but the .666" diameter section was left larger, somewhere around 3/4". The step was turned to tighly accept the displacer shell. The center was drilled 7/16" diameter and parted off a little over finished length.

Reversed in the chuck holding by the 3/4" diameter section, the bore was machined complete with the fine thead for the nut and the end hole was opened out to .470" to clear the displacer rod. Finally,wrung onto a stub mandrel, light cuts took the 3/4" diameter section down to .666" to lighten it. The special nut was made to easily screw into the displacer end thread.

The displacer rod bushing was turned to a close running fit on the displacer rod and a slip fit into the displacer end bore. The length of the bushing equalled the depth of the bore in the displacer end minus about .160", which is the lengh of the nut and the thickness of an o-ring slightly compressed. When the nut is tightened, it pushes the bushing against the o-ring to effect a seal.

---

[3] Chapter 9 of the book *Ringbom Stirling Engines*, op. cit., gives extensive technical details on the modes of stable operation of single cylinder Ringbom engines.

Displacer Assembly

The displacer rod cap simply isolates the rod and its bushing from the insulation placed in the hot end interior of the displacer . This protects against any stray insulation particles wedging between the rod and its bushing and retarding or jamming the displacer. The cap was pushed over the displacer end and secured with retaining compound .

The insulation helps to keep the rod from overheating. The engine has been operated without it, but does run noticeably warmer. It is worth the trouble of putting it in so that the engine can be run at the high temperatures it is capable of. Ceramic fiber insultion was used in the prototype. The assemply drawing shows its type and placement and the means to secure it against the acceleration forces that act on it when running. An alternative that may work adequately is a loose pack of fiber glass, but this has not been tried. After the insulation was in place, the displacer end was pushed in with a light application of sealant . Again, a few tiny pins would give added security. The total mass of the finished displacer was just under 37 grams. This is a fairly high mass. The displacer (and hot end) could be made longer and then only a few convection/radiation shields would be needed inside the displacer in place of the insulation material. This could substantially reduce displacer mass and increase the speed potential of the engine.

Cold End

The cold end was a simple turning from aluminum bar stock. The four holes for the studs had to be accurately located because they aligned the engine parts upon assembly. The fins help keep the engine cool, but are by no means optimal, especially for unaided convective cooling. The two half-round grooves down the bore were made by drilling 1/4" diameter before the main bore was made. This gave an interrupted cut when boring, but light cuts near the end got the job done satisfactorily.

Hot End

A deep drawn stainless steel cup was found for making the hot end on the prototype Thumper. As mentioned, the ID of the cup determined the bore size of the engine. Aternately, the hot end could have been machined from solid or fabricated as discussed for Tapper. The four stud holes in the mounting flange were jig drilled for accurate placement.

Connecting Rod and Piston Yoke Nut

The connecting rod was made with ball bearings at each end for low friction. It was drilled for lighter weight. The bearings were secured with retaining compound. The the piston yoke nut carries the wrist pin for the small end of the connecting rod and secures the displacer rod assembly to the piston. The wristpin is 1/8" diameter . Tubular spacers are fitted to center the connecting rod and keep it from rubbing on the sides of the yoke slot. Soft paper gasket washers were used against each side of the piston crown.

Bearing Support Assembly

The two 1/4" bore ball bearings for the crankshaft are mounted in bearing blocks, each of which is supported on two bearing pillars. The four pillars are secured to the bearing ring with countersunk screws from the underside. This ring fits around the cylinder, rests on top of its flange, and is held in place by the four long studs that reach to the hot end flange and hold the main engine structure together.

Silicone Gaskets

Sandwiched on each side of the cold end are 1/16" thick silicone gaskets. These not only seal the engine capsule, but provide the cushion against which the displacer ears stop. Since no sheet material was available for making the gaskets, some was “cast” from ordinary hardware store variety silicone rubber sealant. The idea was to squeeze out a ring of silicone on a piece of waxed paper, put another piece of waxed paper on top, and squeeze it out between two thick flat plates with 1/16" spacers placed around the periphery. It would have taken a very long time to cure that way with only the outer edge exposed, so an upper plate with a hole was used. The hole was placed roughly over the center of the gasket being formed, and the top waxed paper cut open to allow the inner portion of the gasket ring to communicate with the room air . After a few days the top plate was removed and the gasket left to cure still another day. Then the paper was peeled off and it was cut to shape and the holes punched.

Crankshaft

The crankshaft was made in two removable halves so that disassembly would be easy when making any modifications or adjustments. It has proven to be adequately strong and true running. However, an overhung crank arrangement would be more suited to experimental work since the stroke could then be more easily changed. The displacer is very responsive in this engine. A piston movement of only about 1/8" is enough to cycle the displacer. The engine could therefore be run with a shorter piston stroke, say 7/16" or even 3/8". This would allow the engine to start up and run at lower temperatures, but of course the overdriven speed limit would be correspondingly lower.

Assembly and Operation

The engine was mounted on the stand shown in the photographs for running. Solid alcohol tablets were mainly used for fuel, but a wick or gas burner is necessary for runs of longer duration. Insulation material was put in place around the hot end to keep the upper portion of the engine cool.

Soft sponge rubber pads were glued to the bottom of the stand to cushion the engine from the surface on which it rests. This plays the same role as the resilient motor mounts used in automobiles. Soft springs would also work well for isolating the vibration of this engine.

An added feature not shown in the drawings is a vent screw for the same purpose as on the L-27. This is tapped into the side of the cold end and can be seen in one of the photographs of Thumper.

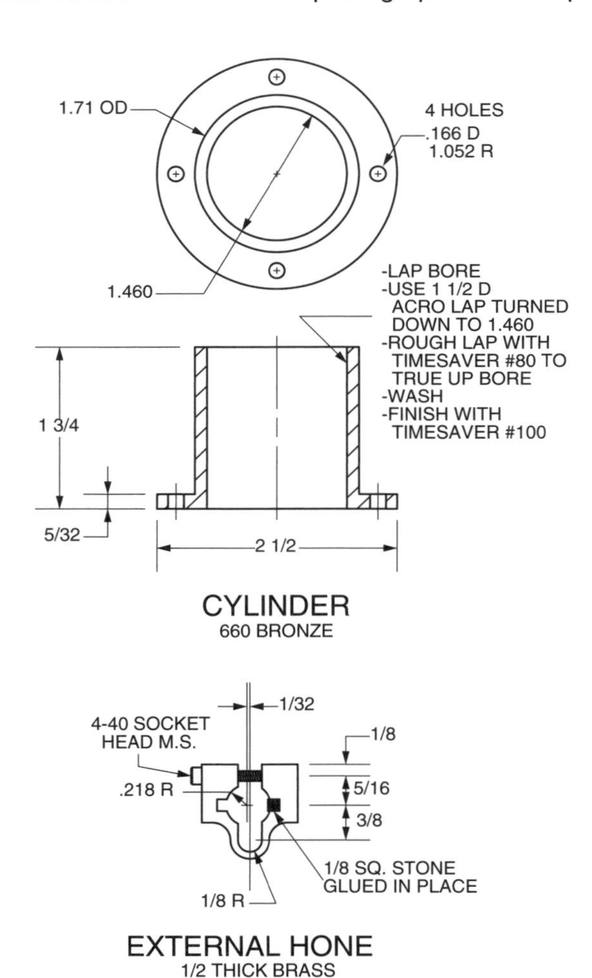

CYLINDER
660 BRONZE

EXTERNAL HONE
1/2 THICK BRASS

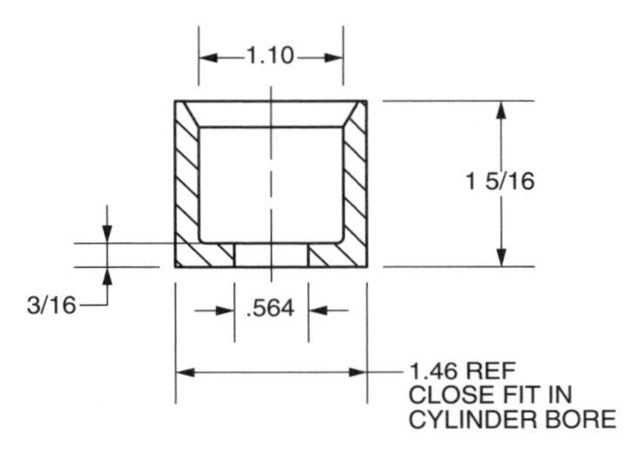

MACHINING SEQUENCE

-SAW OFF BLANK OF SUITABLE LENGTH
-ROUGH OUT INTERIOR AND BORE HOLE
-RECHUCK LIGHTLY AND FACE TOP TO FINISHED LENGTH
-MOUNT ON FACE-MANDREL TO FINISH MANDREL OD SHOULD BE EASY FIT IN ID OF PISTON. CLAMP PISTON TO MANDREL BY SCREW THROUGH CENTER HOLE
-TURN OD TO CLOSE FIT IN CYLINDER

PISTON

GRAPHITE (POCO AXF-5Q)

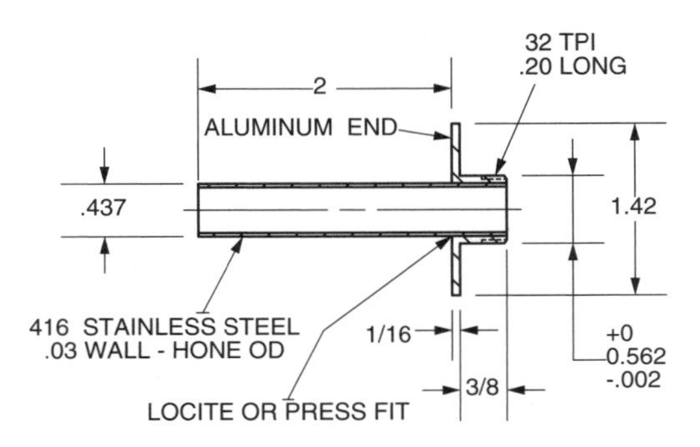

DISPLACER ROD

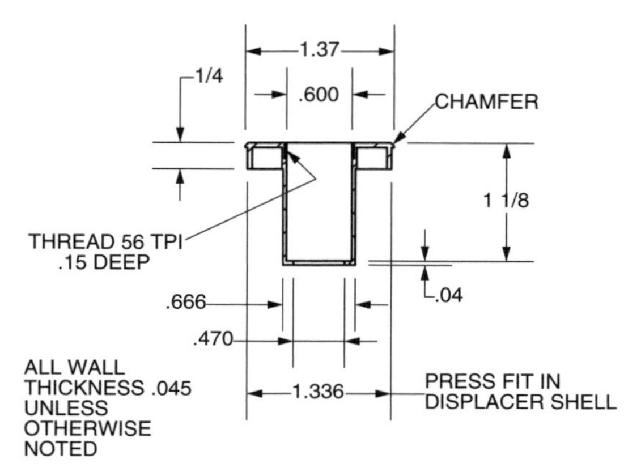

DISPLACER END

ALUMINUM

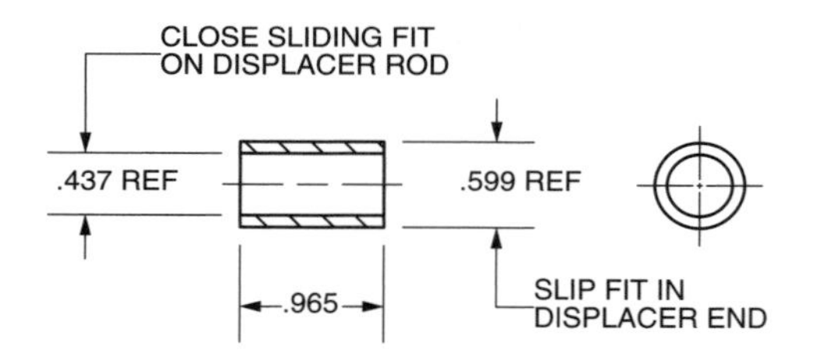

DISPLACER ROD BUSHING

GRAPHITE (POCO AXF-5Q)

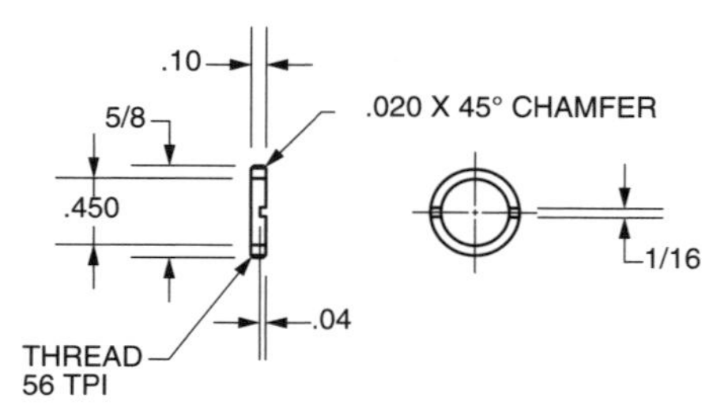

DISPLACER BUSHING NUT

ALUMINUM

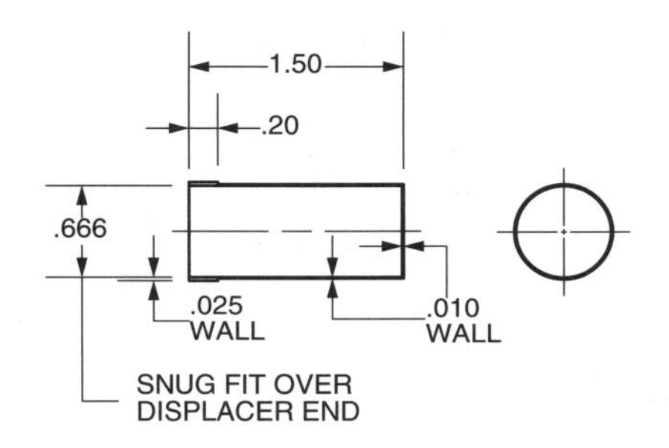

DISPLACER ROD CAP
ALUMINUM

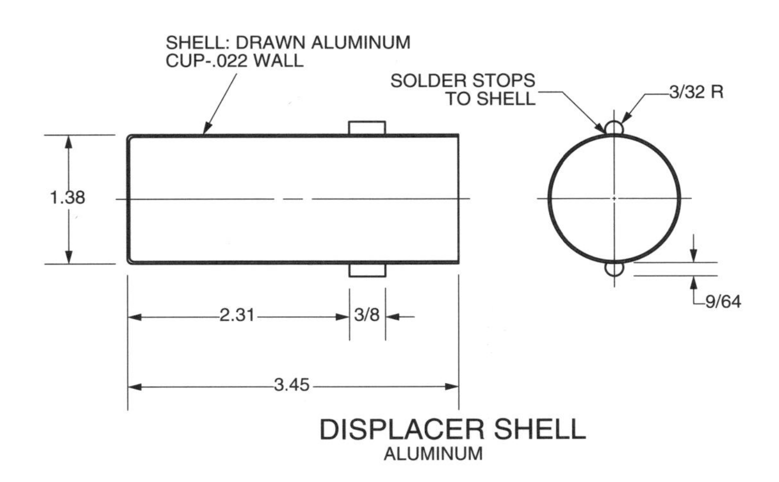

DISPLACER SHELL
ALUMINUM

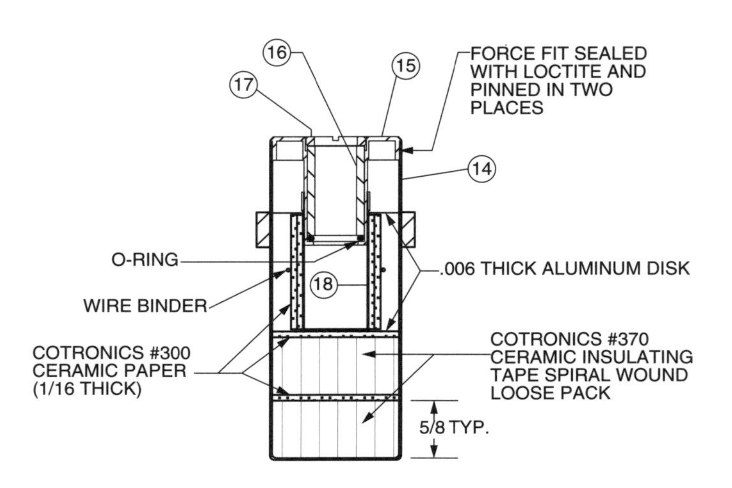

DISPLACER ASSEMBLY

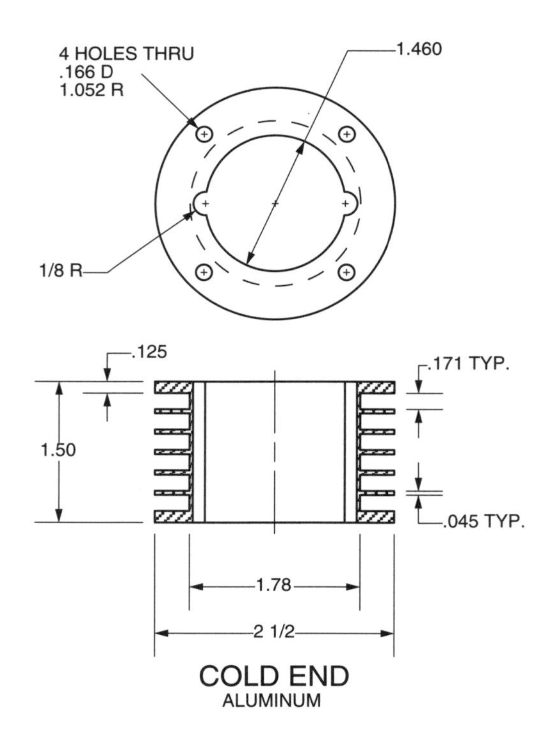

COLD END
ALUMINUM

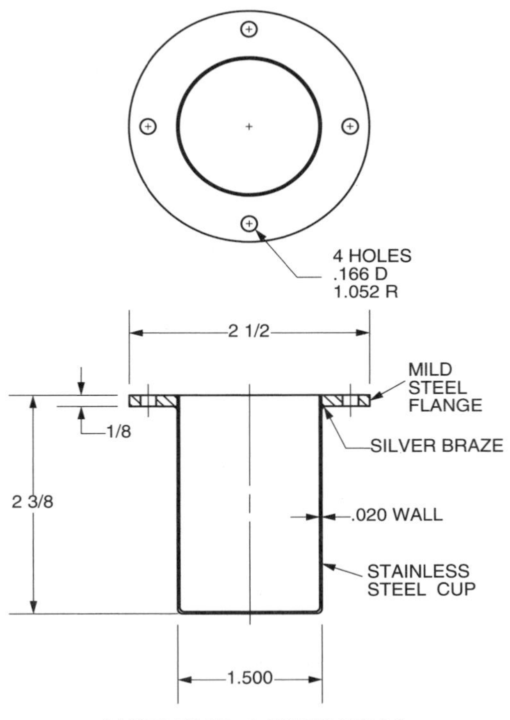

HOT END ASSEMBLY

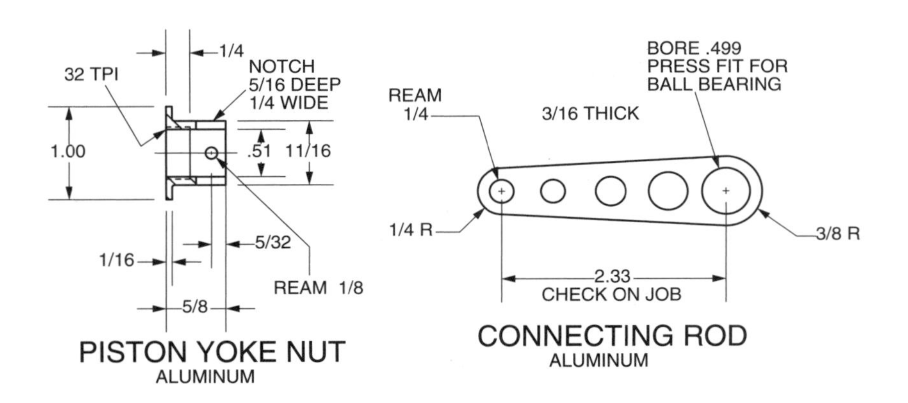

PISTON YOKE NUT
ALUMINUM

CONNECTING ROD
ALUMINUM

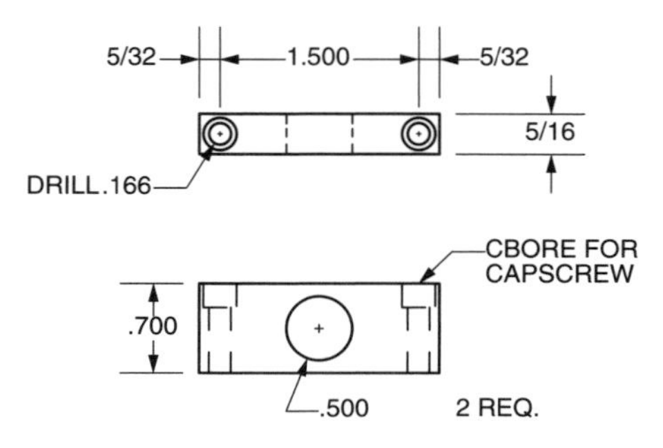

BEARING BLOCK
ALUMINUM

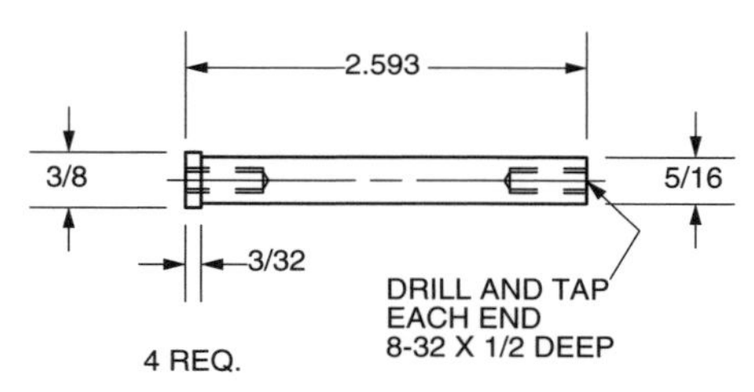

BEARING PILLAR
ALUMINUM

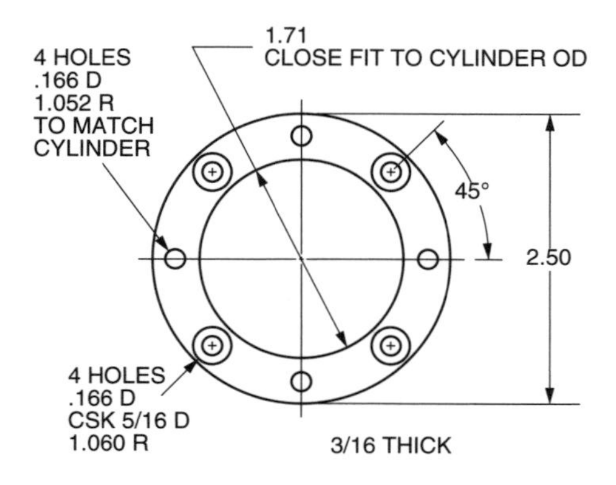

BEARING RING
ALUMINUM

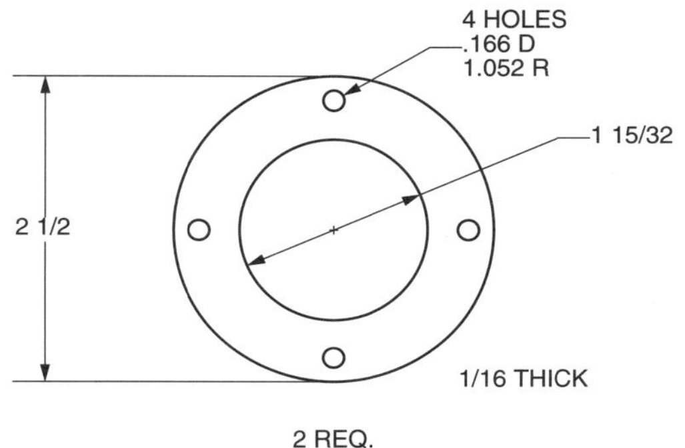

2 REQ.

GASKET
SILICONE RUBBER

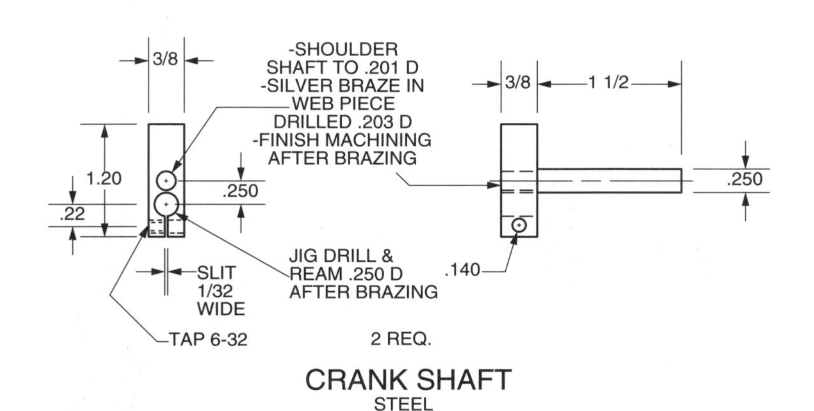

2 REQ.

CRANK SHAFT
STEEL

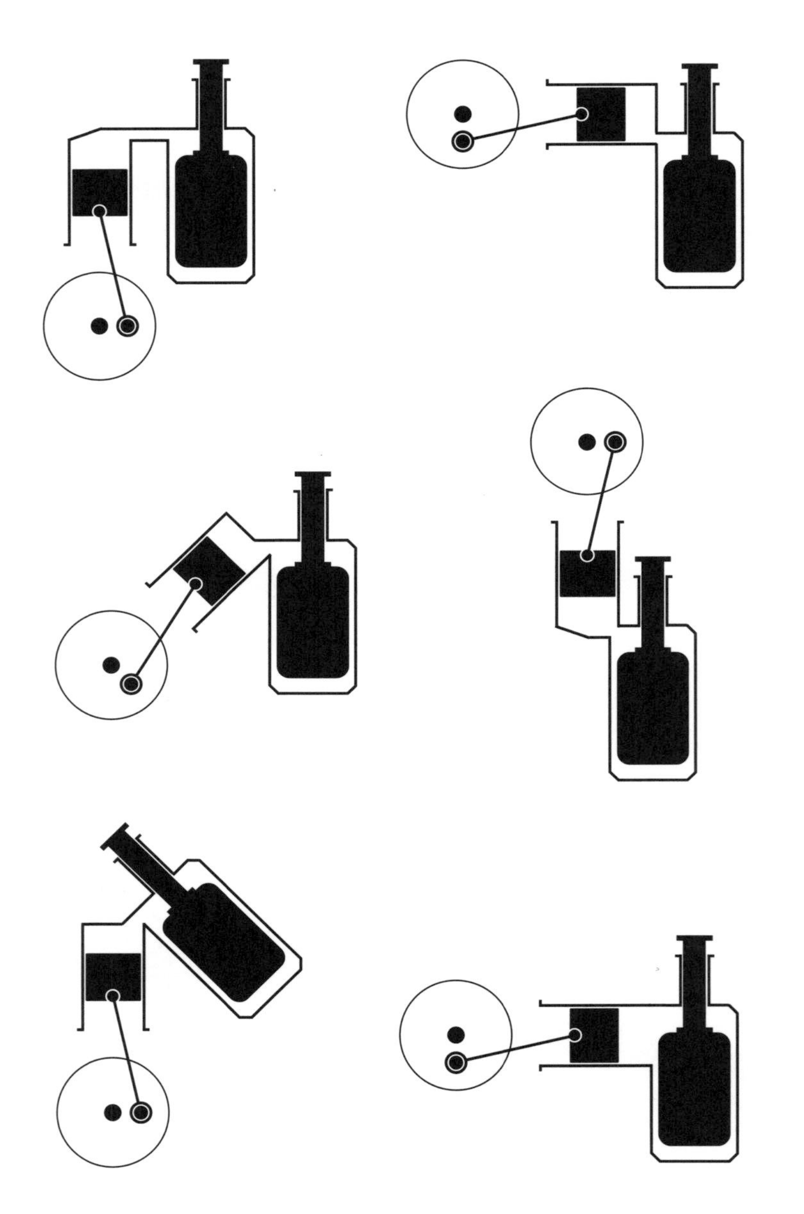

Myriad are the possible configurations for Ringbom engines. These figures show Ringboms arranged with the hot end of the displacer oriented downward.

Chapter 5. New Engine Ideas

Once having made a Ringbom engine more or less to an established design, builders will no doubt come up with many clever ideas for improvements, or maybe even entire new engine designs. The possibilities for new and different Ringbom engines are indeed vast. In this chapter, some ideas are presented for interesting or unusual Ringbom engine variations, along with design guidelines gleaned from mathematical studies of Ringbom engine dynamics.[1]

Ringbom Engine Configurations

Ringbom engines can be configured in all sorts of ways that are practically impossible for conventional kinematically driven displacer engines. The displacer cylinder in a Ringbom can be oriented just about however one wishes. This is not only because the displacer is free of any mechanical connection to the rest of the engine, but also because of the fact that in the modified or overdriven mode Ringbom, gravity is not depended upon for normal operation. Once a well-built engine gets up some speed, it can be turned sideways or upside down without missing a beat.

Thus the displacer cylinder can be mounted hot end up as well as hot end down. The displacer cylinder can also be oriented horizontally as in most conventional model hot air engines. When making a horizontal displacer engine, the displacer rod gland should be proportioned somewhat longer because the displacer is cantilevered from it.[2] A longer gland gives a more favorable "lever of support" and will relieve the friction on the displacer rod. This friction will of course tend to retard the motion of the displacer, but with an adequate gland

---

[1] For an extensive treatise of the mathematics of Ringbom engines, cf. *Ringbom Stirling Engines* , Senft, Oxford University Press, 1993.

[2] This applies to kinematically driven displacer engines as well.

length, a lightweight displacer, and a smooth accurate rod, a horizontal Ringbom has no trouble at all getting up to respectable speeds.

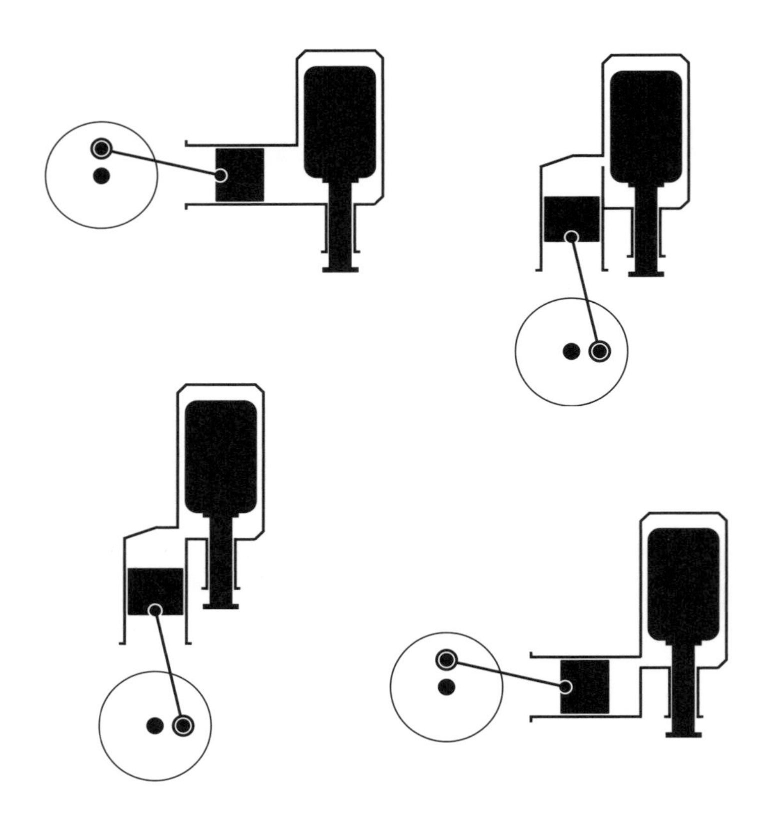

Many hot end downward configurations can be rotated to yield workable engines with the hot end oriented upward. Although more difficult to heat, this arrangement makes it easier to keep the cold end cooler.

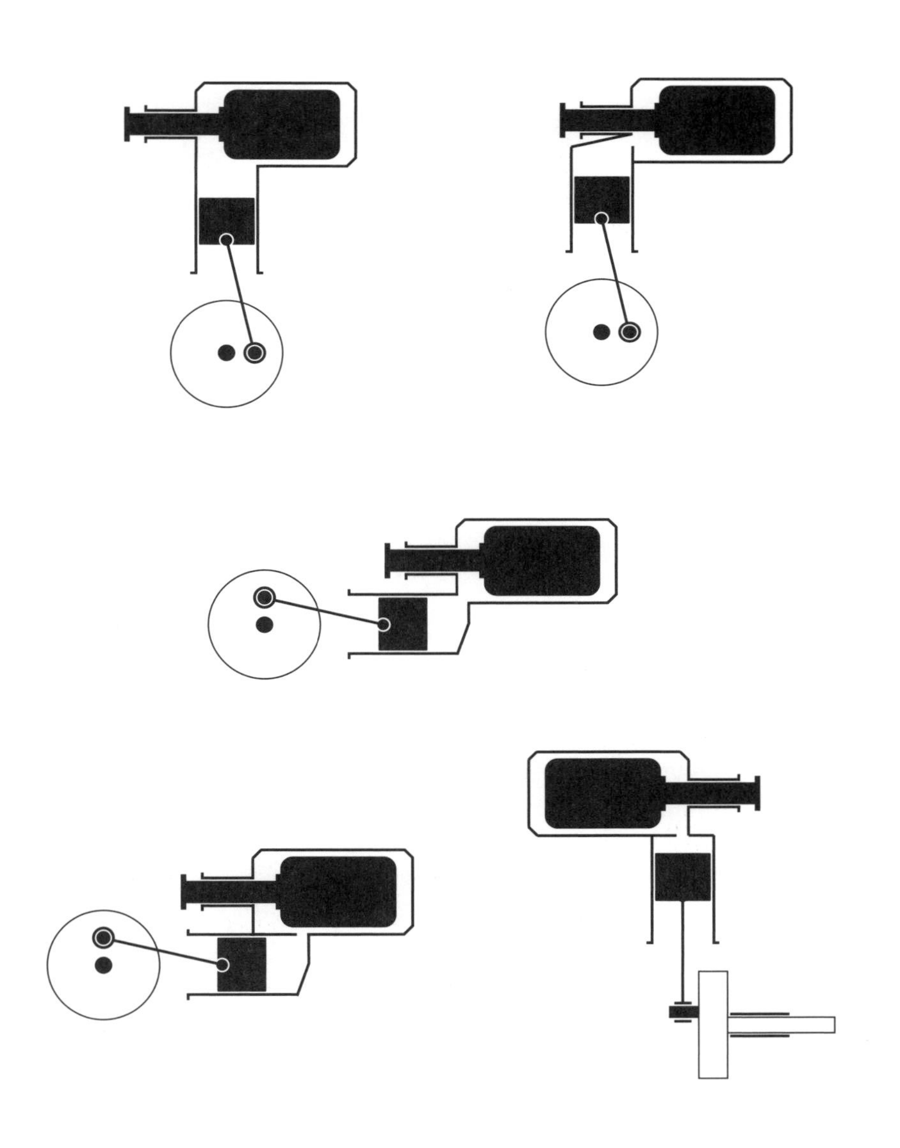

Horizontally oriented displacer units are easier to heat and cool.

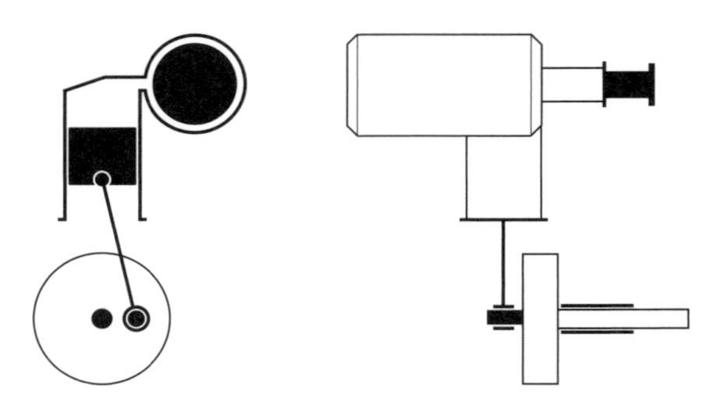

Two views of another Ringbom arrangement in which the displacer unit is horizontal.

However the displacer unit may be positioned, excessively long connecting passages between it and the power cylinder should be avoided. This is because they introduce "dead" volume into the engine. The air in this volume does not take an active part in the cycle but only serves to transmit pressure from one part to another. Thermodynamically, dead volume diminishes the pressure fluctuation the engine generates. This lowers the potential power output of the engine. In low temperature differential engines, this is not very critical because their compression ratio is extremely low by necessity. Just out of curiosity, I have operated the L-27 with its power cylinder unit temporarily connected to the displacer section by a flexible tube several feet long! The speed and power are much lower this way of course, but still the engine runs surprisingly well. It makes a very amusing and dramatic demonstration of the Ringbom mode of operation to be able to pick up the half of the engine with its flywheel merrily spinning away while the displacer unit is still on the table sitting on top of its dish of hot water dutifully bobbing up and down!

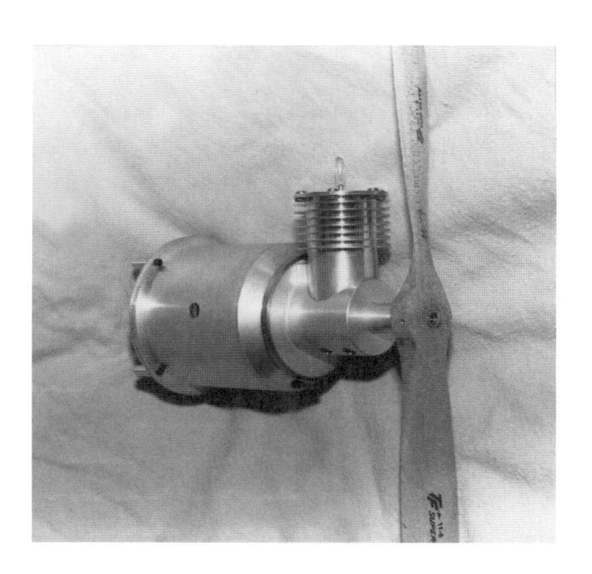

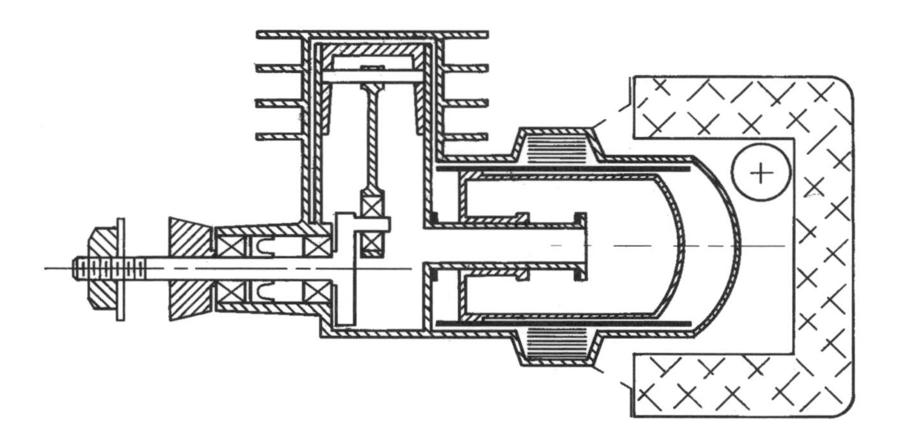

An example which dramatically illustrates the configurational flexibility of the Ringbom is this innovative design by Rob McConaghy. Intended to power a model airplane, the displacer has been positioned behind the crankcase. For compactness, the rod is stationary and works inside the displacer as on Thumper. The crankshaft carries a lip seal so the engine can be pressurized to produce the power density necessary to fly. Surrounded by an insulated firebox, the hot end is fired by a butane burner. The engine is still under development.

Multicylinder Ringboms

The configurational flexibilty of the Ringbom makes it easy to contrive multicylinder engines. The normal restraints of making multiple mechanical connections isn't there, and one can position the displacer ends for more effective heating and cooling.

One could put say three or four or more displacer units in a horzontal radial array with their hot ends facing inward and just about touching one another. Then a single big flame could be used to heat all of them at once. They could even be angled hot end upward , like logs set up for a bonfire. The power cylinder unit could also be made in radial arrangement, located below and connected to the cold ends. This would necessitate a vertical shaft as on a lawn mower, and a flywheel horizontally positioned like a top. Another possibility is to arrange the displacers parallel in a square of four with a wobble plate drive for the pistons. An engine like this would be an incredible sight and sound with all its displacers automatically cycling in and out. If you get busy, you could be the first one to actually see it!

Another interesting variation would be to connect multiple displacer units to just one big piston unit with the same displacement as the sum total the individual units would have. Then all the displacer units would cycle in unison as the engine ran. The advantage of this scheme is greater surface area for heat transfer than a single larger displacer unit would have. More area translates into more rapid heat transfer which makes higher engine speeds possible.

Other good arrangements for the displacer units are in-line or opposed. A very attractive configuration for a double is the one used by the Stirling brothers in their 1843 engine which they built for powering the machinery in the Dundee foundry.[3] Of course, as a Ringbom, the linkage to drive the displacers wouldn't be needed. A difficult aspect of constructing this engine is in the double acting power cylinder. It would not be too easy to make a close fitting

---

[3] Illustrated on p.66 of *Introduction to Stirling Engines*, Senft, Moriya Press, 1993.

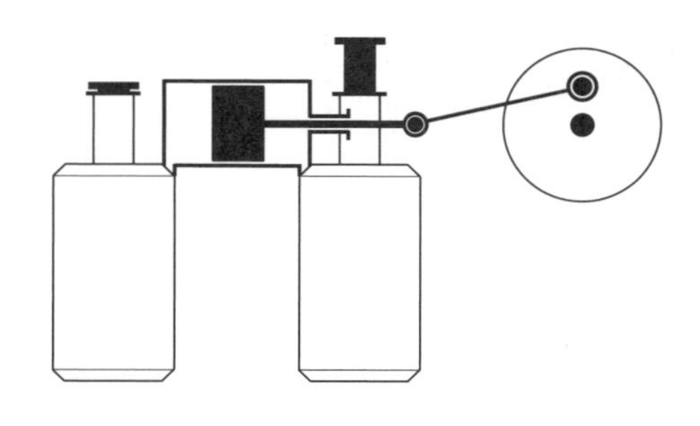

graphite gland and piston perfectly aligned to minimise fricton. Double-acting cylinders work well on steam engines because there is enough pressure to use packing or a piston ring, and these also allow for slight misalignments. But in an atmospheric Stirling, there is not enough driving pressure to afford wasting any power on moving tight seals. Another difficultly is leakage. Piston leakage always has a negative effect in engines, but in this type of Ringbom leakage around the piston would have an exaggerated effect. This is because it would affect the pressure in the adjoining displacer chamber and thus could upset its dynamic relationship to the piston position. Nevertheless, a careful machining job, or a floating type connection between the piston and the rod, might make this idea into a fine running engine. Two ordinary single-acting pistons back to back with a short connecting link is another alternative.

Also to be considered for a twin displacer Ringbom is a pair of single-acting pistons set to run ninety degrees out of phase. Because as explained in Chapter 1 a modified Ringbom will arrive at a mean pressure equal to atmospheric a short time after starting up, the pressure acting on the piston varies from above to below atmospheric. This means the piston is alternately pushed outward and then inward each revolution. So the crankshaft is driven as by a double acting piston. If a second piston is arranged to operate ninety degrees from the first, the engine shaft will always be being turned by one piston or the other, as in a steam locomotive. This requires less of a flywheel and gives a smoother running engine. This is especially important if the engine is put to work driving a load like a Tinker Toy merry-go-round or an Erector Set Ferris wheel.

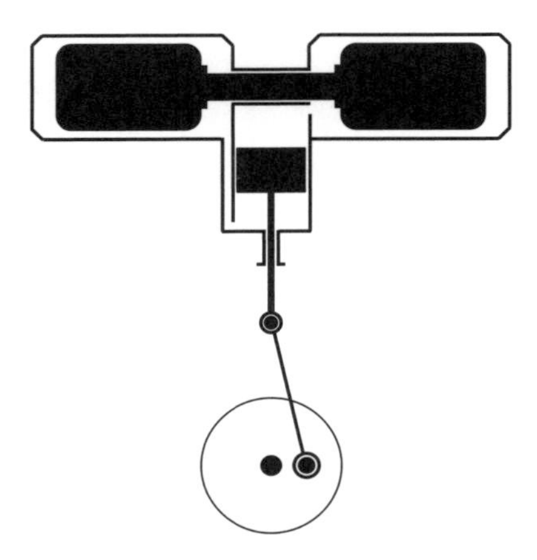

In this double cylinder Ringbom idea, the two displacers are connected to a common rod. The double -acting power cylinder is shown below the displacer rod for clarity, but can be moved upward in a different plane for greater compactness.

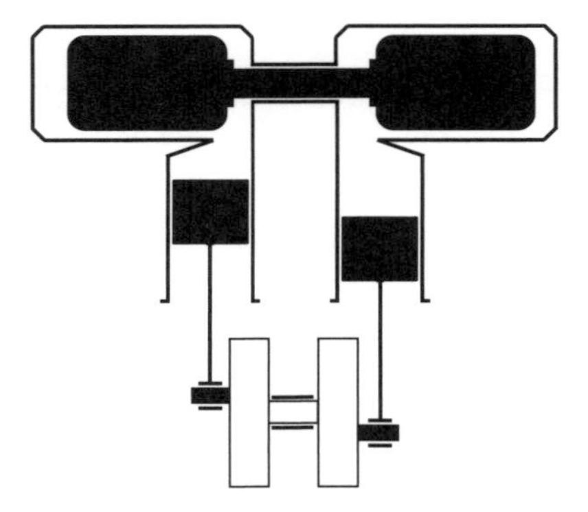

This arrangement idea shows the same opposed displacer unit as above, but connected to two single-acting power cylinders driven by cranks 180° apart. It can also be made more compact as desribed above. Instead of a vertical orientation, these engines can also be set up horizontally taking these figures as a view from above.

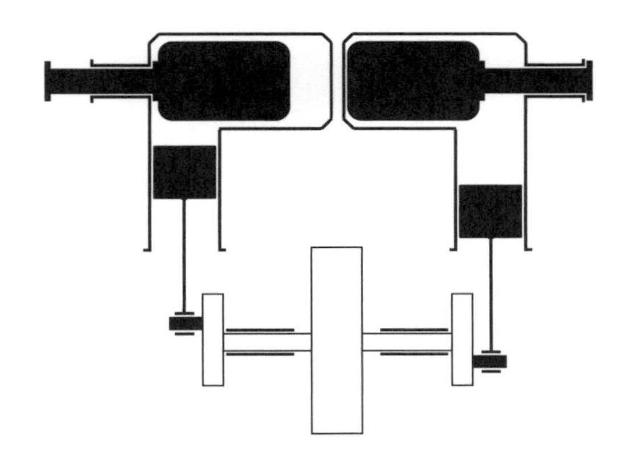

In this double Ringbom concept, the displacer units are arranged with hot ends facing each other, which makes convenient heating from a single source possible. Because the displacers are independent, the cranks can be set at any angle to each other. An angle of 90° will give smoother and almost continuous shaft torque as explained in the text. The cranks could also be set at 180° to give displacer motion in unison. If the cranks were set together, the motion of the displacers would be opposite and balanced. In the latter case, the displacer cylinders could be joined together into one long cylinder heated in the middle and cooled at each end, and the two pistons could be replaced by a single larger one.

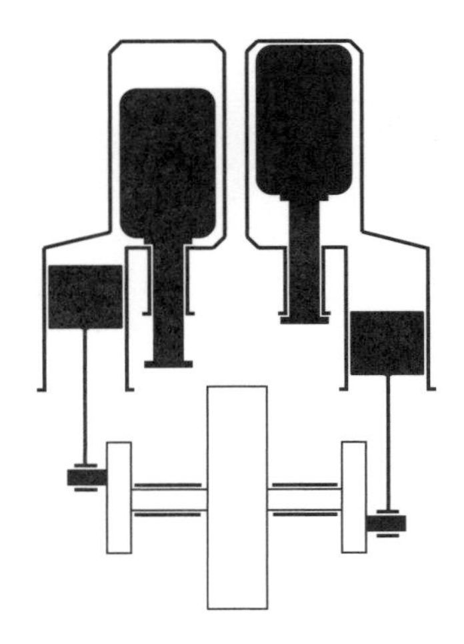

This is a nice simple arrangement for a horizontal twin Ringbom engine. The hot ends are close together so that a single firebox can be used for heating. The cranks can be set at 90° or 180°.

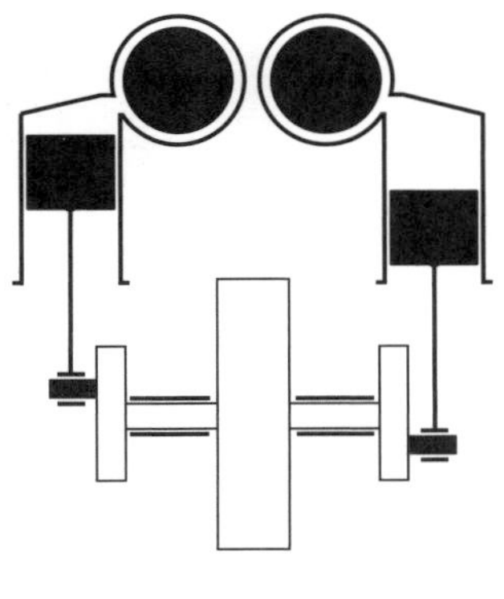

An alternate arrangement for a simple twin Ringbom which could be set up with the displacers horizontal or vertical. In the vertical case, the hot ends could be oriented upward or downward.

Low Temperature Differential Ringboms

The Ringbom makes an ideal low temperature differential (LTD) engine. All engines capable of running at small differentials in temperature, no matter what kind, must have a low compression ratio.[4] For a Stirling this means a large displacer swept volume relative to the piston swept volume. Now the minimum displacer length in a Stirling is roughly dependent on the temperature difference across it. Therefore relatively short displacers are possible for LTD Stirlings. This leads to a pancake-shape displacer with a short stroke. Short displacer strokes are favored for Ringboms because of lower transit times, and a large diameter displacer has a natural dashpot effect as it approaches the flat plate ends of the displacer chamber, another desirable feature for a Ringbom. Finally, LTD engines are intrinsically low speed devices because speed is dependent upon heat transfer, and heat transfer is driven by temperature differential. This favors intermittent rapid displacer strokes with long dwell periods, a prime characteristic of the Ringbom. For these reasons, Ringboms are ideally suited for low temperature differential operation. If you have already built your own L-27, you well know that.

The author has produced two other LTD Ringboms, one bigger and the other smaller than the L-27. Actually, the larger one, called the ANL,[5] was made before the L-27, being the first ever LTD Ringbom. The displacer of the ANL engine is about 8-1/2" in diameter and the maximum piston swept volume is about 9 cu.in. (150cc). This engine, in addition to being an effective demonstrator of Ringbom and general heat engine characteristics, proved to be an excellent laboratory engine. Much was learned from the ANL engine. A very vivid lesson learned was that regenerators dramatically improve the performance of LTD engines. It was found that the peak power of the engine near-

---

[4] The relationship between operating temperatures and compression ratio is described in detail in the paper "Mechanical efficiency of kinematic heat engines" , Senft, *Journal of the Franklin Institute*, 1987, Vol. 324, p.273-290.

[5] ...because it was designed by the author while a visiting scientist at Argonne National Laboratory.

ly doubled with a regenerator incorporated into its displacer. [6]

Another more subtle lesson learned in the course of designing the ANL engine is that the source temperature operating range for LTD Ringboms is limited above as well as below. This is easily observable when operating a LTD Ringbom on a small quantity of warm water. As the engine runs and the water cools, the engine runs slower and slower and finally stops of course. If the piston stroke is then reduced, which is equivalent to lowering the compression ratio, the engine will again run on the cooler water. This is common behavior for all LTD Stirlings. But for a Ringbom, if the source is then replaced with piping hot water, the compression ratio may be set too low and its displacer will just lock at one extreme or the other. To run again the piston stroke must be increased. The lower the compression ratio of a Ringbom engine, the more limited is its operating temperature range.[7] This is why it is important to have a way of varying the piston stroke on a LTD Ringbom if you wish to operate it over a wide range of source temperatures. The ANL engine operated down to a temperature differential of 13°F (7°C) when its piston stroke was reduced to 30cc.

The L-27 was designed largely by scaling down the ANL in a way that preserved or enhanced its best qualities. Likewise, a smaller engine, named the L-77, was later also designed by scaling. This engine has a displacer just under 3" in diameter and a piston swept volume of 4cc. Like the L-27, the regenerator frontal area of this engine is about 30% of the total displacer area, but is incorporated in a different way. In the L-77, the regenerator is in the form of an annulus of polyurethane foam glued around the periphery of the styrofoam disc. This was simple to make once a uniform strip of the regenerator foam was cut. The same arrangement could also be used on the L-27. The little L-77 on top of a hot cup of coffee runs at a terrific clip.

---

[6] Many results of tests done on the ANL engine, including pressure volume diagrams, are reported in Chapter 8 of *Ringbom Stirling Engines,* op. cit.

[7] For a rigorous explanation of temperature limits in Ringboms cf. *ibid.* p.119-121.

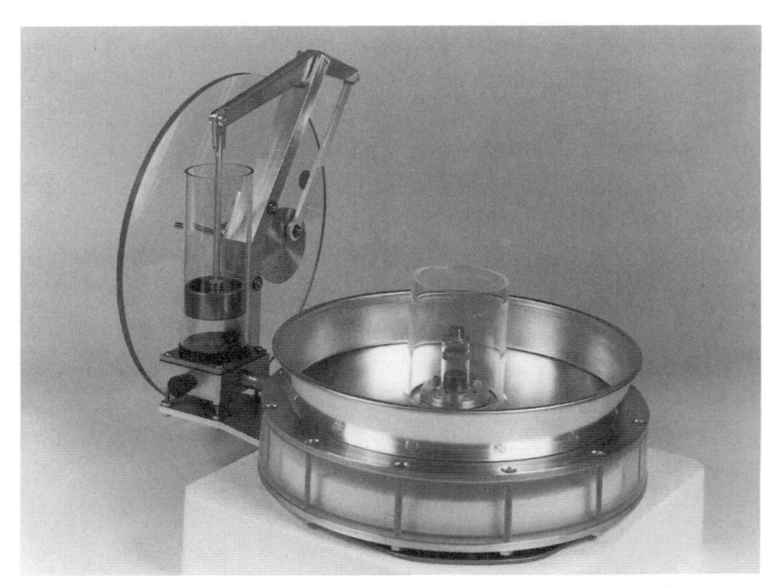

Designed and built in 1983, the ANU engine was the first Ringbom LTD engine.

The smallest member of the ANU / L-27 family is the L-77 .

Scaling Theory

Designing a Ringbom engine from scratch is not an easy matter. The dynamic behavior of Ringboms is sufficiently complex that mere human intuition isn't enough to decide right off what proportions to choose. Mathematical analysis is a great aid in designing Ringboms. The mathematically inclined reader should enjoy delving into the full set of equations that describe Ringbom engine behavior, but this is only really necessary when starting a completely new design.[8] More often than not, one only wishes to take an existing engine and make another one pretty much like it, but bigger or smaller. As mentioned, the ANL engine was scaled down into the L-27, and then farther down into the L-77. Simple scaling is easy. The hard work is in coming up with a good first design.

Simple scaling actually requires only the arithmetic of multiplying all of the original engine dimensions by a common scaling factor. Of course, a scaled engine will run differently than the original, but the difference is simple and predictable for Stirlings. Furthermore, *simple scaling of a Ringbom will affect its performance the same way that simple scaling a conventional kinematic Stirling will affect its performance.*

One should first understand how simple scaling affects the general performance of any kind of Stirling engine. Suppose all the dimensions of a given engine are to be multipled by the factor α . For the sake of illustration, first suppose the engine is being made larger, that is, the scaling factor is greater than one. Engine volume increases by the cube of the scaling factor, namely by α^3. The heat transfer surface area however increases by the square of the scaling factor, namely by α^2. Now the rate of heat flow into, and out of, the engine is proportional to surface area. If the same cycle is to be carried out in the larger engine, the energy input per unit volume will have to be the same. Since the energy transferred is the product of rate and time,

---

[8] Ibid., Chapter 3.

the time required for each cycle will have to be multiplied by α. This means engine speed is divided by the scaling factor.

This turns out to be a pretty good rule of thumb. If you double the size of an engine, it will run at about one-half the speed, and produce about four times the power. If you triple the size of a hot air engine, without changing its form or features, it will have a top speed about one-third and nine times the power of the original. Scaling the linear dimensions of a Stirling engine by the factor α multiplies its speed by $1/\alpha$ and its power by α^2. If speed is what you are after, reducing size is an easy way to get it. According to this relation, if an engine is scaled to one-fourth of its original size, it is capable of running four times as fast , and its power will be one-sixteenth of the larger engine.

This same rule is obeyed by Ringboms. Of course, they are Stirlings too, but what the astute reader is wondering here is whether the overdriven speed limit would scale the same way. It is not obvious, but it in fact does scale the same way. To see this consider what we will call the Ringbom Comparison Equation: [9]

$$\frac{\bar{\varpi}_2}{\bar{\varpi}_1} = \sqrt{\frac{A_2\ L_1\ M_{D1}\ p_{c2}}{A_1\ L_2\ M_{D2}\ p_{c1}}}$$

This equation applies to two Ringboms, engine #1 and engine #2, that have the same temperature extremes to work between, have the same ratio of rod area to displacer cross sectional area, and the same ratio of displacer swept volume to piston swept volume. On the left hand side is the ratio of the maximum overdriven speed, $\bar{\varpi}$, for each engine; the subscripts refer to engine #1 and engine #2. On the right hand side is the square root of a ratio of engine parameters: A is cross sectional displacer area; L is displacer stroke; M_D is displacer mass, and p_c is the mean pressure of the engine, which is usually atmospheric pressure.

Now consider a simple scaling of all linear dimensions by the fac-

---

[9] For a derivation of this formula, cf. ibid., Chapter 6.

tor α; say engine #2 is scaled from #1. Then L_2 will be α times L_1 and A_2 will be α^2 times A_1. Usually the displacer length is not scaled exactly because it depends considerably on the temperature difference the engine works between, but as a general approximation let us assume all dimensions are treated the same. Then the mass M_{D2} of the new displacer will be α^3 times M_{D1}. If the charge pressure of both engines is the same, then with a little algebra the comparison equation shows

$$\frac{\bar{\omega}_2}{\bar{\omega}_1} = \sqrt{\frac{(\alpha^2 A_1)\, L_1\, M_{D1}\, p_{c1}}{A_1(\alpha L_1)(\alpha^3 M_{D1})\, p_{c1}}} = \sqrt{\frac{1}{\alpha^2}} = \frac{1}{\alpha}$$

This is exactly the same speed multiplier as that obtained from heat transfer considerations, so Ringboms are not restricted any more than is any other kind of Stirling engine.

Changing Displacer Proportions

If it is desired to increase the speed range of a given Ringbom, the Comparison Equation is a good guide to doing it effectively. Suppose the displacer diameter is again scaled up by the factor α. The equation as mentioned is only valid if the rod to displacer area ratio is the same for both engines being compared. Therefore suppose the rod diameter will be increased by the factor α also. Keeping the same piston swept volume in the new engine requires that the new displacer stroke be scaled down by the factor $1/\alpha^2$ to keep the displacer swept volume the same. Now the displacer length need not be increased in this instance, so the displacer mass increases roughly by the factor α^2. Thus

$$L_2 = (1/\alpha^2)L_1 \quad \text{and} \quad A_2 = \alpha^2 A_1 \quad \text{and} \quad M_{D2} = \alpha^2 M_{D1}\ .$$

Putting these values in the Comparison Equation gives

$$\frac{\bar{\omega}_2}{\bar{\omega}_1} = \sqrt{\frac{(\alpha^2 A_1)\, L_1\, M_{D1}}{A_1 (1/\alpha^2) L_1 (\alpha^2 M_{D1})}} = \sqrt{\alpha^2} = \alpha$$

which shows the overdriven speed increases by the factor α.

As an example, if the displacer unit of a Ringbom is replaced by a new one having a displacer and rod 50% bigger in diameter (this means $\alpha = 1.5$) and a stroke which is about 44% of the old one ($1/\alpha^2 = 4/9 = 0.44...$) then the modified engine will be capable of running a full 50% faster. Of course, limited heat transfer or friction may prevent the engine from actually reaching this full speed potential, but the engine is dynamically capable of a 50% higher speed. This discussion gives a good demonstration of why Ringbom engines favor large short-stroke displacers rather than longer stroke displacers.

Pressurization

Another way of increasing the speed and power potential of a Ringbom is by pressurizing it. The Comparison Equation given above shows that Ringbom engines respond extremely well to pressurization.

The reader is no doubt aware that pumping more air into any kind of Stirling engine is like supercharging an internal combustion engine. The engine, although it is still the same size, has more air to work with, like a bigger engine would have, so it performs like a bigger engine would. The result is a corresponding increase in power, everything being perfect. For example, if you were to double the mean pressure in a Stirling, it would have twice the amount of air inside of it to work with, and so could make twice the power. It is like two engines compressed into one. That is the basic idea behind pressurizing or "charging" as it is also called.

The scaling equation above shows how the top speed limit of a Ringbom is affected by charging. If the mean pressure p_{c1} is raised by the factor h to become $p_{c2} = h\ p_{c1}$, but all other variables are unchanged, then the ratio of the new overdriven speed limit ϖ_2 to the old ϖ_1 is given by

$$\frac{\bar{\varpi}_2}{\bar{\varpi}_1} = \sqrt{\frac{p_{c2}}{p_{c1}}} = \sqrt{\frac{(h\ p_{c1})}{p_{c1}}} = \sqrt{h}$$

This means for example, that if the engine is charged to double the pressure, the new overdriven limit will be $\sqrt{2} = 1.41...$ times the old limit. This is an increase of 41%. The more the pressure is raised, the higher will be the overdriven speed limit, everything else being the same.

It must be understood that before any Stirling can be pressurized, it must be equipped with an enclosed pressure-worthy crankcase or "buffer space". This is necessary to minimize bearing loading and preserve the mechanical efficiency of the engine. Without a pressurized buffer space, a typical simple Stirling engine will just stop when more air is pumped into its working space.[10] A buffer space in effect surrounds the entire engine with an artificial atmosphere of denser air in which to work. Its work per cycle increases proportionally, and usually its speed also increases due to improved heat transfer.

A major technical problem associated with pressurization is getting the power out of the engine. Usually this is done via a dynamic seal on the engine shaft.[11] Typical are lip type seals made of Teflon or face type seals of graphite. Keeping seal friction in check is essential to obtain the benefits that charging promises. An alternative is to enclose the entire shaft and flywheel and extract the power by a magnetic coupling working through the crankcase wall. The new rare earth "super" magnets would be ideal for making simple couplings of this sort. It is important to use a material for the crankcase that has zero,or at least acceptably low, eddy current loss.

However a Ringbom may be enclosed in its own pressurized atmosphere, the displacer rod must be included. This is automatic in a beta-type engine. In a gamma or split-cylinder type like Tapper or L-27, fixing a cap over the displacer rod is all that is needed. This gives the displacer rod an enclosed space in which to work, a little

---

[10] For more technical information on the subject of charging, cf. the following paper: "Pressurization effects in kinematic heat engines" Senft, *Journal of the Franklin Institute*, 1991, Vol. 328, p.255-279.

[11] For more information on seals, cf. e.g. the classic book *Stirling Engines* by G. Walker, Clarendon Press, Oxford, 1980.

"crankcase" as it were for the rod. This space will gradually come to the same mean pressure as the rest of the engine through leakage past the rod and so the entire engine will be operating inside its own uniform elevated atmosphere. If the space provided is small, a "spring effect" will be introduced on the displacer; this can often improve operation as is discussed below.

Novel Components

There is plenty of scope for productive tinkering with Ringboms. Being a fairly recent arrival on the model engineering scene, much has yet to be tried for the first time in these engines. Described here are a few ideas for alternative or additional components for Ringbom engines.

Displacers. The common canister-type displacer tends to always turn out heavier than one wishes. Normally the internal pressure of a displacer equals the mean cycle pressure. The engine cycle pressure fluctuates above and below this. Because the pressure outside the displacer goes above the inside, the displacer wall needs to be heavier than if the pressure were always higher inside. Thus one way to reduce displacer thickness is to inflate or pressurize the displacer interior at or above the peak engine pressure.

A simpler alternative to this "balloon type" displacer is to use the style of displacer that Laubereau introduced on his hot air engine of 1861.[12] It has no interior, its longitudinal section being in the form of an "H". This also permits extended heat transfer area on both the hot and the cold ends.

Diaphragms. The precision required by pistons and cylinders in model engines makes one contemplate alternatives such as a liquid piston or a diaphragm. Diaphragms are attractive of course primarily

[12] This engine is illustrated on p. 64 of *The Evolution of the Heat Engine*, Ivo Kolin, Moriya Press, 1998.

because they are easy to make and suffer no leakage. However, they do suffer from internal friction resulting from the distension of the diaphragm material. This can amount to a relatively large loss for a low power engine. Nevertheless, diaphragms have been successfully used by several builders. Foremost among these is Prof. Ivo Kolin who pioneered the use of diaphragms on all of his LTD Stirlings.

A diaphragm can be used in place of the power cylinder unit in a straightforward fashion. One only need match the swept volume of the diaphragm to the cylinder it replaces. But because diaphragms are large-area by short-stroke devices, they generally cannot be used directly to drive the displacer in a Ringbom. However, a small lever arrangement can be employed to amplify the stroke of the diaphragm. Again, the swept volume of the displacer-driving diaphragm should equal that of the displacer rod being replaced.[13] The lever then needs to be proportioned to produce the desired displacer stroke from the diaphragm stroke.

Displacer Springs. There are two kinds of springs that one can attach to the displacer of a Ringbom to improve performance. One is a so-called centering spring used on engines where the displacer moves vertically. This is a very soft spring[14] whose purpose is to just counter the weight of the displacer and suspend it in a more or less central positon when the engine is idle. Just hanging there, the displacer readily responds to the slightest movement of the piston which makes it easlier for the engine to get itself going.

The second kind of displacer spring is a stiffer sort which acts in both directions, compression and extension, and is set up so that it is relaxed when the displacer is at its midstroke position. The spring is not stiff enough to prevent the displacer from completing its strokes or to eliminate dwell periods. It is stiff enough however to store some energy when compressed or extended which materially aids the acceleration of the displacer when it is ready to start moving in the

---

13 ...which is cross-sectional rod area times displacer stroke.
14 ...i.e. a spring with a small constant.

other direction. Done correctly, a spring like this will increase the overdriven speed limit of the engine.[15] The spring most conveniently takes the form of a gas spring, rather than a mechanical spring.

Displacer Dashpots. Ossian Ringbom described dashpot devices to rapidly bring the displacer to a halt as it neared the end of its stroke. These took the form of a collar-and-socket. A short collar was located at each end of the displacer rod. At each end of the gland, a recess or counterbore was cut just slightly larger than the rod collar diameter. As the displacer neared the very end of its stroke, its collar would enter its mating socket forming a pneumatic cushion or dashpot to rapidly decelerate and stop the displacer.

Soft rubber cushions seem to serve well enough for displacer stops in small model engines. For better performance, one could make stops from one of the special foam elastomers presently available which are specially designed to dissipate impact energy. In LTD engines, the large flat displacer has an excellent dashpot effect as it nears its flat plate ends, so nothing additional is needed.

In large engines running at fair speeds, some form of non-contact stop is essential for long engine life. Collar-and-socket dashpots as Ringbom described are easy to make and hard to beat. An improvement to try would be the incorporation of one-way valving into the socket so that the displacer would not be retarded as it attempted to leave when beginning its next stroke.

Stroke Limiters. It is rarely required in model Stirlings to control engine speed and power. Actually, it is hard to do in conventional kinematic hot air engines anyway. It is by contrast very easy to do in a model Ringbom. All one need do is provide a device which would move one of the displacer stops. Moving the outer one closer to the other would cut down the stroke of the displacer and this would reduce engine power. It would make an interesting and instructive feature to include on a model Ringbom.

---

[15] For a thorough discussion of displacer springing, see Chapter 7 of *Ringbom Stirling Engines*, op. cit.

Although built as an end in itself, this beautiful Ringbom engine hardly qualifies as a miniature! In fact it is almost as tall as its builder, Olaf Berge, who wanted to create an engine much like the one Ossian Ringbom himself might have made back in 1907. Mr. Berge fabricated the engine from scratch by cutting, bending, welding, and machining. The author assisted from a distance only by carrying out some calculations. These calcualtions showed what the engine performance would be for various sizes of the main components. When a satisfactory combination was found, construction went ahead. The displacer diameter is 6" with a stroke of 3". The displacer rod diameter is 2-1/4". The cylinder bore is 4" with a piston stroke of 5-1/2". The engine is shown here at rest, with its displacer assembly all the way down, and therefore only its end is visible. The rod assembly incorporates collar-and-socket dashpots made with lip seals to work only in the desired direction. These are very effective in quickly but smoothly decelerating the displacer at the ends of its strokes with no bumps or knocks. The engine is a fine runner, quietly operating stably in excess of 200 rpm, a speed which compares well with other hot air engines of the same era.

Father,
may everything we do
begin with your inspiration
and continue with your saving help.
Let our work find its origin in you
and through you reach completion.

Liturgy of the Hours • Morning Prayer • Monday • Week I